Samuel Hough Terry

Controlling Sex in Generation

Samuel Hough Terry

Controlling Sex in Generation

ISBN/EAN: 9783337371074

Printed in Europe, USA, Canada, Australia, Japan

Cover: Foto ©berggeist007 / pixelio.de

More available books at **www.hansebooks.com**

CONTROLLING SEX

IN

GENERATION.

THE PHYSICAL LAW INFLUENCING SEX IN THE EMBRYO OF MAN AND BRUTE,

AND

ITS DIRECTION TO PRODUCE MALE OR FEMALE OFFSPRING AT WILL.

WITH AN APPENDIX OF CORROBORATIVE PROOFS,

By SAMUEL H. TERRY.

THIRD EDITION,

REVISED AND CORRECTED BY THE AUTHOR

NEW YORK:

CLARK & ZUGALLA,

33-43 GOLD STREET.

1889.

PREFACE

TO THE SECOND EDITION.

IN the evolution of a theory of the action of any hitherto unknown law of nature there are two stages :

The first is, after a sufficient number of relevant facts are collected, to deduce from them some general principle which will fit them all. When this is found, the idea takes the position of a hypothesis, more or less probable as the facts considered have been many in number and sufficiently varied.

The second stage is, when well-authenticated facts derived from other persons can be brought to corroborate the hypothesis, especially such facts as were gathered for some different purpose. When these are sufficient to satisfy the honest intelligent reader the evolution of the idea should be regarded as complete, and what was before but a hypothesis be permitted to take rank as a theory.

Naturally when the investigator arrives at the end of the first stage *he* is convinced. To *him* the deductions and conclusions he has made from the facts in his possession seem so complete and convincing that he thinks they should be quite apparent to others on short reflection.

The influence of these convictions must be my justi-
fication for publishing the first edition with no more
of corroborative proof than was inserted in its pages.
And the desire to fully corroborate my views the
excuse, if any be needed, for the additional matter
in the Appendix of this edition.

THE AUTHOR.

NEW YORK, February 1887.

CONTENTS.

(iii)

CHAPTER VII.

CHAPTER VIII.

CHAPTER IX.

CHAPTER X.

CHAPTER XI.

APPENDIX.

CHAPTER I.

THE writer of this little treatise has been a husband and a father over thirty years. The first five children his wife bore him were girls. A natural desire for sons led him to make some investigation into the subject of the origin of sex, and the first movement was to collect facts that seemed to bear on it. After much consideration of these facts, and of the constitutional characteristics of those families where a tendency to the production of one or the other sex exclusively prevailed, he formed a theory on the subject, and, being the owner of some farm stock, he commenced experiments with them in the line of this idea, by which he was led gradually along to the adoption of the theory herein presented as the true physical law governing the reproduction of sex, and which he hesitated not to put in practice in his own marital relations, with the result that of the last four children born to him three were boys.

That in every instance the reduction of this theory to practice will produce as certain and definite results, will not now be positively claimed. The peculiarities of our physical condition are so varied, and often mysterious, that it is well known that what will act on one person, is inert on another. Remedies which are accounted specifics in certain diseases occasionally fail. And it is enough to claim a like general rule for the operation of the " physical law " herein presented.

This much is done with confidence, so far as the comparatively limited range of one person's experience can give confidence. By the nature of the subject there can be but limited opportunities of comparing the experiences of other persons in practically testing the theory so as to give greater confirmation. The theme and the instructions are such as modesty shrinks from introducing as subjects of ordinary conversation in the social circle, and it is only in the form of a printed volume that an opportunity can be given to others to test the theory.

It is hoped that confirmations of the theory in the personal experiences of any reader of this book will be communicated to the author through the publishers, as also failures where the conditions herein named are attempted to be complied with, giving the physical traits and other proper information regarding the parents in both cases. Such communications tend to confirm the truth. and to establish an array of facts all-important in further adapting the theory to the varied phases of man's physical condition. Such communications will be held reverently sacred and confidential, and, in any public use of them, the names of parties, and everything that can direct attention to the persons from whom they may come, will be withheld.

Perhaps a few words may be appropriate here on the propriety of making public such a discovery. There are a great many really good people who regard every new discovery in the laws of nature, by which man is able to control what has before been regarded as the special action of an overruling Divine Providence, as a blasphemous and sinful thwarting of God's will. Rather than use direct means to obtain definite ends, they prefer uncertain results, believing that thus the Great Disposer of all events decides for them, and, of course, His decisions are always right.

" Where ignorance is bliss 'tis folly to be wise," and I would not by an unnecessary act disturb the trusting faith of any. But since the Garden of Eden was closed against our first parents, and the decree went forth that men must labor for their daily support—must toil and struggle with the " thorns and briers" which the visible world everywhere brings forth, and in the sweat of their brows must earn their bread—everything that tends to ameliorate their hard lot must have God's and every good man's commendation. The workers of our race are necessarily the men. It is these that are specially fitted to the task. To the women is given the work of continuing the race. In " pain and sorrow are they to bring forth " till the end of time. It is the man's duty to care for her—to provide food, clothing, and shelter, she aiding, of course, incidentally, but without neglecting the great duty of her life.

To this end God set the race apart in families, appointing unto one man one woman. In the natural order of events, that this equal division should prevail, many more sons should be born than daughters. The greater loss of life among the former, owing to their greater exposure to dangers, would leave the sexes at a marriageable age nearly balanced in numbers. But what is the condition now ? For a half century past the women have been gradually increasing in proportional numbers to the men.

The following table, made up from the last two United States Census Reports, shows the relative number of males and females composing the populations of the principal Eastern and Middle States of this country, being those where this surplus of females is greatest :

STATE.	1870.			1880.		
	Males.	Females.	Surplus females.	Males.	Females.	Surplus females.
New Hampshire.	155,640	162,660	7,020	170,526	176,465	5,939
Massachusetts	703,779	753,572	49,793	858,440	924,645	66,2·5
Connecticut.....	265,270	272,184	6,814	305,782	316,918	11,136
Rhode Island....	104,756	112,597	7,841	133,030	143,501	10,471
New York ..	2,163,229	2,219,530	56,301	2,505,322	2,577,549	72,227
Pennsylvania ...	1,758,499	1,763,452	4,953	2,136,655	2,146,236	9,581
New Jersey......	419,672	456,424	6,752	559,922	571,194	11,272
Maryland.	384,984	395,910	10,934	462,187	472,756	10,569
	5,985,829	6,136,329	150,408	7,181,864	7,329,264	197,400

	Increase in 10 years.	In population.	In female surplus.
State of Massachusetts		23.5 per cent.	33 per cent.
" New York............		16 "	30 "
In the 8 States named.........		11 "	31 "

The Census of 1880 shows that in the States east of the Alleghany ridge, there were over

1*

300,000 more females than males then living, fitted, or soon to be fitted by nature and by education, to become the married partners of a li number of the opposite sex, and in th: ' the rejoicing mothers of happy families, who have to forego all the felicities of married life, and all the enjoyment of seeing sons and daughters growing up around them, simply and solely because there are not men in their vicinage for them to mate with. Without homes of their own, and without the near kindred ties and the solace of husband and children, they are condemned, for no fault of their own, to pass through life aimless and unhappy because failing to accomplish the great end of their existence as women, too often regarding life itself as a burden, and looking forward to the grave with contentment as their only refuge.

Reader, does the thought of this bring merely a smile to your face, or, worse, a sneer, with the usual shallow remarks about old maids? Ought it not bring tears rather? In the whole circle of unfortunates in the world at this time, there are none more deserving of heartfelt sympathy than these isolated and lonely superabundant women, many of them the very loveliest of their sex.

Nothing tends more to detract from the value of anything in the eyes of mankind than a surplus of it. This is a general truth applicable to ....en as well as to ordinary earthly possessions. A thing is raised in our estimation if we find a difficulty in obtaining it, even though the necessity for it be not great. Printed calicoes at ten cents a yard are too mean and common for any but the poor to wear, but when they were raised in value by the high price of cotton during the late war to fifty cents a yard, how handsome they looked! fitted to adorn the persons of those who aspired to be considered wealthy. This being a governing influence in valuing lesser things, how does it affect the men in search of wives, who, looking around the circle of their acquaintance, see twice as many unmarried women of a marriageable age as there are men? Is not their own importance in the human family as men likely to be overrated, and their appreciation of the opposite sex (some of them perhaps their own sisters) much depreciated? Will they think it necessary to be uniformly moral and circumspect in their conduct to ensure their obtaining respectable and virtuous women for their wives? Notoriously this is not regarded as necessary now.

When the man's own consequence in the marital condition is thus so unduly exalted, what prospect of happiness can there be after marriage for educated and refined women—such as have been taught that God created the race, male and female, of one blood to dwell in equality together? especially for such of them as are unwilling to sink below their proper level without some struggle. Society is fast becoming converted to the idea that the bride must bring with her to the altar a certain number of dollars, as a dowry, to make her equal to the husband.

It is unnecessary to pursue the argument further. It must be clear to every one who reflects on the subject, that the large surplus of women in the community is now the fruitful cause of many social evils as well as moral ones, and that this surplus is year by year increasing. In view of this, shall there be any doubt of the propriety of publishing to the world a modest theory that professes to show how this relative disproportion of the sexes originates, and how the equilibrium can be restored? To state the case is to decide it, and I will not insult the moral perception of my readers by any further remarks on the subject, but will pass on to the preliminary considerations appropriate to the purpose of the treatise, trusting that even if the reader, when he finishes it, disagrees with me in my inferences or conclusions, he will test them by his own observations and experiences before he finally decides against them ; and that he will at least believe that I have been actuated by an earnest desire to do good in publishing the work, and that I honestly believe in the truth of the theory advanced.

CHAPTER II.

THIS subject of controlling the sex of offspring
in generation is not one limited to the mere nar-
row desires and wishes of the family, as might
be thought by those who simply glance at the
subject. The whole of society, not only in this
country, but throughout the civilized world, must
on reflection deem it of vital interest. As it is
important that the reader have a sufficient conception of the magnitude of this subject, that he
may take a larger view of it, than simply as it
affects him individually, some labor has been
bestowed in gathering and collating statistics
tending to show not only the increasing proportion of women, but how and why it exists.

An opinion somewhat prevails that this increased proportion of women is mainly confined
to our ·Eastern States, and is due principally to
the migration of the young men from this locality to the Western and Pacific States. While

(14)

this increases the disparity, it is insufficient to account for the larger part. For it must be remembered that we have also a large immigration from foreign countries to these Atlantic States, which is so largely composed of men as to be considered the principal reason for the increased proportion of women in the European countries from which the emigrants came; England and Wales alone showing a surplus of over 500,000 females at the last census.

The statistics of births are so imperfectly kept in this country that it is not possible to fully prove from them what the author alleges, and will endeavor to substantiate—which is, that the tendency of mothers who live in luxury and ease is to the production of female offspring. It needs but a careful observation around the circle of any one's city acquaintance to see that in many of the older families in our towns, where there have been two or three generations of those living in indolence, that among the more recent descendants there is not an average of one boy born to two girls. While the statistics do not on their face, and in the aggregate given, show this condition, deductions drawn from them prove it to be true in many places.

It may fairly be inferred also from the statis
tics given that this result is not from any condition of vitiated air, or other unhealthy characteristics often thought to be specially unfavorable
to vitality in large cities, for this condition of
more numerous female births occurs among the
well-to-do people of the country towns, and seemingly wherever there is found in a family so
much wealth that for two generations the wives
and daughters are enabled to live in ease, and
therefore too often without sufficient muscular
exercise to give full vigor to their bodies.

The first result of this languid and inactive
life is to produce a laxity and tenderness of the
muscular system. We are wont to call this *effeminacy*, and the word is appropriate and sig-
nificant, for under its influence the race tends
largely to the production of females.

Strange as it may seem, and not in accordance
with popular belief in the relative powers of en-
durance in the two sexes, the statistics show conclusively that the boy in the first two years of
existence more readily succumbs to disease than
the girl, and it is only a fair inference that the
same result follows in the various stages of the
fœtal existence. Indeed the records, on subse-

quent pages, of still-born children, both in New York City and in Philadelphia, showing very uniformly through a number of years—in the former 1,000 boys to 682 girls, and in the latter 1,000 boys to 712 girls—may be taken as quite conclusive on this point.

This seeming anomaly is not without corroboration in the vegetable world. It is found by a special cultivation which tends to delay and enfeeble the growth of the plant, some varieties of strawberries can be made to produce only or mainly the female blossoms, and by a higher cultivation and more stimulating plant food be made on the other hand, to produce mainly male blossoms. Thus showing that in the enfeebling of the plant the male characteristic is the first to succumb.

A circumstance noticed by some recent Arctic explorers corroborates this idea. They mention finding the female variety of a bisexual plant many miles further north than they found the male variety; the inference being that the female had better withstood the rigors of the inhospitable climate—had continued to passively live, where the male, if ever able to exist there, had died out.

Conversing on this subject with an intelligent gentleman—Mr. Nelson Sizer, Vice-President of the American Institute of Phrenology, and a member of the Fowler & Wells Co., publishers of this book—he suggested a very natural cause for this. As he said, " The females of all creation are endowed with an ability to assimilate food beyond their own individual needs, for the supply of their offspring ; placed in straitened circumstances, especially when this additional supply is not called for by the offspring, it would add to their ability to maintain their hold on life."

There is a tendency in the vegetable world for the parent plant to throw off or abort whatever fruit it can not bring to perfection, as seen in the premature falling off of the surplus settings of fruit on a tree at their very early period of formation. It is a reasonable inference that this characteristic is so fundamental that the enfeebled parent sometimes refuses to nourish the germ it is incapable of perfecting at its very incipiency, and that similar influences are exerted in the case of enfeebled mothers in the animal creation.

The United States census of 1880 shows that the great proportion of the surplus females is composed of those between the ages of 20 and 30, being over one-third of the whole.

In order to show that this surplus of females of marriageable age is consequent largely on the decreased number of boys born in the effeminate conditions mentioned, and still more largely upon the increased number of boys dying in infancy from the, as I allege, same conditions, I present some statistics of births and deaths from different localities embracing a fair average of the years when these now grown people were born. From which I draw the inferences and make the deductions in proof of the allegations.

These vital statistics will, I believe, be found more reliable for other purposes than those compiled in the years following the civil war; it having been often noticed that when society is unsettled, and business affairs uncertain, as they were for a time after the war, there are not so many marriages, and the same prudential motives would tend to decrease the number of births at such times. Besides this the very large immigration to this country since the war would tend to make the statistics of births unreliable as giving the normal results from our own people.

STATISTICS OF BIRTHS IN MASSACHUSETTS.

	1849.	1850.	1851.	1852.	1853.	1854.
Boys............	13,329	14,137	14,949	15,246	15,798	16,352
Girls.	12,263	13,392	13,613	14,432	14,965	15,469
Proportion :						
Boys	1,000	1,000	1,000	1,000	1,000	1,000
Girls........	920	947	911	947	947	946

The details by counties are given for only one year, that of 1854, a fair average year, in which is also given the nativity of the mothers.

COUNTIES IN MASSACHUSETTS.	Births.		Parentage of Mothers.	
	Boys.	Girls.	Native.	Foreign.
Barnstable	399	384	692	86
Berkshire	592	612	782	384
Bristol	1,090	1,096	1,395	768
Dukes.................	56	37	83	4
Essex	2,173	2,038	2,581	1,355
Franklin.............	416	333	514	105
Hampden	759	709	793	557
Hampshire	448	421	585	237
Middlesex	2,841	2,661	2,792	2,463
Nantucket.............	62	56	112	10
Norfolk................	1,489	1,471	1,493	1,414
Plymouth	769	822	1,211	318
Suffolk	3,137	2,899	1,934	3,949
Worcester	2,121	1,930	2,207	1,513
Total............	16,352	15,469	17,173	13,163

It may be accepted as a fact, that the foreign-born mothers of the above table are mainly of the laboring classes, not doubting but that a large majority of the native-born mothers are also, it is yet fair to claim that the foreign born more nearly represent the working class, those who by some daily toil and bodily exertion harden their muscular development, and that if a separate classification of such were made they would show a largely increased percentage of male births over the more sedentary and sybaritic class of mothers. Though this can not be definitely proven from the statistics, it may be indirectly. The State at large gives a total of 946 girls born to each 1,000 boys. Of the 26 principal towns in the State having the greatest number of births, 13 or one-half the number have an excess of foreign-born mothers; in proportion 1,843 foreign born to each 1,000 native born. The births of these show to each 1,000 boys 950 girls. They are the following towns:

	Births.		Parentage.	
	Boys.	Girls.	Native.	Foreign.
Cambridge..	336	333	230	420
Lowell.................	564	521	460	580
Roxbury...............	283	270	183	368
Worcester	354	388	333	408
Lawrence	252	231	159	319
Fall River.	157	155	95	216
Dorchester	196	174	174	180
Chicopee	156	119	106	167
Milford.	139	147	98	187
Lee...................	50	55	50	54
Taunton	173	190	150	209
Salem.................	270	256	120	244
Boston	2,945	2,742	1,725	3,806
	5,875	5,581	3,883	7,158

The other 13 towns where the native mothers are in excess,—in proportion 1,000 native to 550 foreign, the births are to each 1,000 boys 975 girls. These towns are:

	Births.		Parentage.	
	Boys.	Girls.	Native.	Foreign.
Charlestown	390	381	428	338
New Bedford.............	248	244	364	119
Lynn.....................	255	256	338	154
Newburyport	158	147	232	73
Springfield	221	201	225	161
Chelsea..................	176	142	187	134
Danvers.................	175	132	174	131
Gloucester	148	159	233	72
Haverhill...............	105	126	169	58
Adams..................	84	98	125	55
Great Barrington	40	47	64	22
Pittsfield	117	123	122	112
Northampton............	102	107	112	93
	2,219	2,163	2,773	1,522

The proof may be further strengthened by taking the extreme cases in each classification; for instance, the six towns in the last list, New Bedford, Newburyport, Gloucester, Haverhill, Adams, and Great Barrington, where the native mothers most preponderate,—in proportion 1,000 native to 336 foreign, and there were born to each 1,000 boys 1,049 girls.

While in the four towns in the previous table,

where the foreign mothers are most heavily in excess, being in proportion 2,120 foreign to 1,000 native, there were born to each 1,000 boys only 903 girls.

That this result is not due to the character of nationality is proven by a summary of the State at large outside of these 26 towns, which will comprise the rural population, and naturally mostly made up of an actively working class. The foreign mothers are there only 426 to each 1,000 native, while the births are to each 1,000 boys but 935 girls.

The statistics of births for the city of New York until recent years are very deficient. Those published as recently as the year 1864 give a gross total of only 5,877 births, while the statistics of deaths show 25,645 for the same year. The Registrar, in his report, estimates the total number of births at fully 32,000 for 1864. Any inference or argument, therefore, drawn from a classification of less than one-fifth of the whole must be weak. Of the whole number returned, 3,059 are boys, 2,818 girls, or, in proportion to each 1,000 boys, 921 girls; being a less proportion of girls than is shown among the rural population of Massachusetts. The parentage is not classified as in that State.

With a view to learn something of the social condition of the parents, permission was obtained to look over the monthly reports as received in the office. Unmistakable evidence abounded in these that the births reported were mainly those among the German population, whose physicians and midwives, having been trained in the country of their birth to the importance of these reports, had presented them with commendable regularity. In fact, with the exception of the reports from the public institutions of the city (and one other exception in which the names were Irish), they all gave undoubted evidence of German origin in the family names, in the chirography, and in the idiomatic errors of language common to that nationality when using the English language, and not proficient in it. We may, therefore, fairly assume that this record gives the proportions of the sexes as they occur in the births among the immigrant population alone.

The statistics of births in the city of Philadelphia twenty years ago, the only other locality in this country from which it has been convenient to obtain them, make an exhibit for four and a half years, as follows :

	Boys.	Girls.	Proportion of girls to each 1,000 boys.
1860 (6 months).	4,426	4,008	906
1861............	9,008	8,263	917
1862............	7,609	7,132	937
1863	8,042	7,251	902
1864	8,237	7,354	893
Totals........	37,322	34,008	Average 911

The nationality of the mothers is not given; but, judging from the statistics of marriages there for the same period, the proportion of foreign mothers was between 700 and 800 to the 1,000 native: probably about the same as in the State of Massachusetts. The proportion of female births, it will be noticed, was less there than in the whole State of Massachusetts, and closely approximates to that of the four towns mentioned, where the foreign-born mothers were in greatest excess. As is well known, Philadelphia city has many rural characteristics—a greater number than is usual in such large cities. The mothers of the families, in a large majority of cases, do their own household work in their own isolated homes spreading out over a large territory, and thus secure a full share of bodily health and vigor.

It needs but a comparatively small percentage

2

of the mothers living in indolent softness, to produce enough girls to materially change the proportion of the sexes; and it may fairly be deduced from the statistics that it is not so much
owing to any greater vigor or robustness in the
great majority of the mothers in Philadelphia,
to which the larger proportion of boys is due, as
to the smaller number who are deficient in sexual vigor as compared with the other localities.

The statistics from all these places clearly show
that, taking all classes, there are more boys born
than girls. It is claimed only that under certain
local circumstances there are increasing tendencies to an undue proportion of girls in the births.
And that the mothers in circumstances of enervation and relaxation of muscular fibre, the results mainly of idleness and luxury, are declining in their ability to produce boys, and that
even those they do bring forth are lacking in
vital stamina.[1]

This first material indication of a decline in
the mother's strength is shown in the proportionally increased number of deaths of boys in
early infancy from debility, marasmus, etc., who,
as before mentioned, seem to succumb first or
more readily to disease.

1 See Appendix, Note A., page 159.

The second stage of declining strength is shown in the increased proportion of boys among the still-born children. The third stage of the decline is a decrease of the conception of boys, more girls being born proportionally from this class of mothers.

This last I have endeavored to show by the statistics already presented.

The second stage is shown by the records of the still-born, given in detail on ensuing pages. The aggregate being in New York City, for three years, in the proportion, 683 girls to each 1,000 boys. In Philadelphia, for the four and a half years, in the proportion, 712 girls to each 1,000 boys. The first stage is shown by the terrible catalogue of infant mortality presented here.

The following mortuary statistics of children under two years of age in the city of New York are for nine successive years, from 1856 to 1864, inclusive, and are so arranged as to show at a glance the totals of each class of organs affected by the disease causing death :

	1856 Boys	1856 Girls	1857 Boys	1857 Girls	1858 Boys	1858 Girls	1859 Boys	1859 Girls	1860 Boys	1860 Girls	1861 Boys	1861 Girls	1862 Boys	1862 Girls	1863 Boys	1863 Girls	1864 Boys	1864 Girls
ORGANS OF RESPIRATION.																		
Bronchitis	71	73	115	89	121	107	97	89	105	91	150	126	107	92	131	106	140	106
Congestion of Lungs	72	54	87	74	91	71	59	48	85	58	74	55	85	63	*	*	*	*
Consumption	104	95	122	80	96	78	85	88	90	61	95	82	104	86	97	90	95	97
Inflammation Lungs	225	180	311	227	310	242	322	246	316	296	233	239	283	215	401	330	414	311
Croup	134	114	156	139	118	110	135	103	150	124	124	67	150	141	203	179	176	148
Diphtheria	*	*	*	*	*	*	13	16	97	83	101	88	134	117	218	215	158	163
	606	516	791	609	736	648	714	590	843	713	777	657	863	714	1,053	920	983	845
ORGANS OF NUTRITION.																		
Cholera Infantum	670	652	640	596	752	759	702	589	569	555	571	578	651	584	650	609	635	590
Diarrhœa	221	176	205	156	185	214	170	174	191	128	123	135	175	135	237	236	209	224
Dysentery	99	92	68	65	89	68	66	54	39	58	48	49	33	27	46	33	80	67
Inflammation Bowels	47	51	78	52	62	53	64	40	67	48	91	43	98	72	90	70	138	95
	1,037	971	991	869	1,088	1,098	1,002	856	806	789	833	805	937	821	1,029	939	1,062	956
ORGANS OF SENSATION.																		
Congestion of Brain	99	67	115	94	97	99	94	85	101	80	85	56	96	52	123	76	87	67
Convulsions	691	561	744	634	776	740	822	646	704	614	657	554	629	555	724	638	687	563
Dropsy of Head	334	287	413	281	426	313	343	284	310	267	347	259	301	229	317	262	303	250
Inflamma't'n of Brain	83	75	110	93	138	87	114	99	118	106	143	109	148	108	166	134	186	115
	1,217	990	1,382	1,091	1,437	1,239	1,373	1,134	1,233	1,057	1,232	978	1,174	944	1,331	1,110	1,263	1,028

CONTAGIOUS DISEASES.																		
Scarlet Fever	105	148	117	134	129	140	203	200	289	285	93	112	97	111	196	208	201	195
Whooping-cough	66	43	40	50	103	88	68	48	81	68	137	130	144	116	95	93	95	71
Measles	67	80	52	64	28	46	107	110	68	54	67	84	96	107	70	98	87	95
Small-pox	64	66	16	9	50	67	103	137	60	64	11	15	118	122	101	110	99	85
	302 (90)	337	225 (87)	257	310 (92)	341	481 (97)	495	496 (105)	471	308 (90)	341	455 (100)	456	462 (90)	509	482 (107)	449
FEEBLE VITALITY.																		
Premature Births	96	142	108	156	124	147	139	176	107	139	152	222	193	204	229	230	145	242
Still-born											+	+	642	856	620	938	611	945
Debility	162	188	102	109	131	142	114	129	137	149	128	140	157	192	159	193	134	192
Teething	37	48	44	44	32	38	60	75	48	64	101	91	72	69	130	173	159	185
Malformation	10	19	17	21	16	22	17	16	6	15	15	17	30	42	36	48	92	42
Scrofula	18	13	13	15	14	19	19	19	27	23	24	31	31	25	26	33	21	26
Marasmus	525	603	573	647	612	567	515	635	574	633	560	661	6-3	698	617	730	577	654
	843 (84)	1,018	857 (86)	994	829 (89)	935	864 (82)	1,050	899 (86)	1,023	980 (84)	1,162	1,707 (911)	2,086	1,817 (84)	2,365	1,6-2	2,358
Deaths from causes not included in the above	204	383	415	471	306	398	422	551	359	437	402	558	395	420	375	361	346	316
TOTAL	4,278 (85)	5,041	4,466 (67)	5,128	3,984 (86)	4,648	4,247 (86)	4,938	4,313 (89)	4,813	4,270 (83)	5,150	5,592 (924)	6,223	5,223 (86)	6,989	4,987 (84)	5,543

NOTE.—The small figures under the footings of the columns of girls give the proportion of these to each 100 boys.

* For the years 1863 and 1861 the deaths by "Congestion of the Lungs" are not given. They are probably included in the list of "Inflammation of the Lungs."

† The record of still-born is not continued after the year 1858.

‡ These figures show the proportions, leaving out the still-born.

It will well repay the reader to give the fore-going record a careful study. It presents more clearly than those hereafter given, the dread effect of certain classes of diseases on the male offspring of our race in the very early period of their existence. Recurring first to the still-born, which for the three years shows to each 100 boys, respectively, only 63, 66, and 75 girls.

The next bearing most heavily on the boys are diseases of the brain—the organ of sensation, averaging, as will be seen, to each 100 boys only 82 girls.

The third in order is the class of diseases indicating feeble vitality, as marasmus, debility, etc., averaging to each 100 boys only 85 girls.

The fourth (and only so because less productive of deaths than the previous) is the class of the respiratory organs, averaging the same as the third, to each 100 boys 85 girls.

The fifth class is the diseases of the nutritive organs—the stomach and bowels, averaging to each 100 boys only 91 girls.

The sixth and last class, the contagious diseases, are equally fatal to both sexes; the deaths from these being about the same as the average proportions of births, that is, to each 100 boys 95 girls.

There is a notable circumstance in the special tendency of some diseases to find their larger proportion of victims from one sex, as dropsy in the head to boys, and whooping-cough to girls. One disease not sufficiently important in the number of cases to be given in the table under a separate head, shows such a remarkable fatality among boys that I present it as an additional argument to show that sex has some special susceptibility to fatal disease. It is jaundice, which for the nine years presents a record as follows :

	1st.	2d.	3d.	4th.	5th.	6th.	7th.	8th.	9th.	Total.
Boys	18	14	14	11	7	..	8	10	11	93
Girls	6	9	7	6	3		5	8	4	48

The mortuary statistics of Philadelphia city for the four and a half years herein presented tell the same story of the increased mortality of boys, though the reports which give the sex are not made up so as to show this feature in its full proportion. These reports classify the deaths of all under 20 years as infants; though a separate classification is made of deaths by the various ages, in which the sexes are not separated.

Statistics of Deaths in the city of Philadelphia for the four and a half years mentioned, of Infants—under 20 Years of Age giving Sex.

	6 months of 1860		1861.		1862.		1863.		1864.		Totals.	
	Males	Fem.	Males	Fem.	Males	Fem.	Males	Fem.	Males	Fem.	Males.	Fem's.
ORGANS OF RESPIRATION.												
Bronchitis	10	12	42	86	37	29	26	31	50	42	167	150
Diphtheria	97	106	248	241	132	169	208	210	155	179	840	905
Congestion of the Lungs	16	9	38	31	53	40	45	30	53	49	205	179
Consumption	42	72	136	166	124	109	119	132	131	175	542	714
Inflammation of the Lungs	56	56	262	222	241	241	234	204	282	193	1,075	916
Croup	94	49	153	150	134	119	245	198	213	235	830	751
	315	304	879	846	721	767	869	825	894	873	3,668	3,615 *98
ORGANS OF NUTRITION.												
Cholera Infantum	244	212	316	302	321	308	456	474	331	310	1,668	1,606
Diarrhœa	29	31	66	48	65	52	63	65	99	64	322	261
Dysentery	50	38	51	40	43	31	51	42	84	56	279	207
Inflammation of the Bowels	51	42	62	64	72	73	67	56	73	55	325	290
	374	323	495	454	501	464	637	637	587	485	2,594	2,363 *91
ORGANS OF SENSATION.												
Diseases of the Brain	20	16	47	34	28	35	41	27	50	28	726	639
Congestion of the Brain	45	43	93	79	103	116	156	132	143	129		
Convulsions	115	108	325	281	330	324	334	294	316	346	1,503	1,813

											Total (M)	Total (F)
Dropsy of the Head	68	65	132	86	98	99	130	92	115	98	677	535
Effusion of the Brain	13	16	42	24	32	24	16	19	31	82	789	698
Inflammation of the Brain	70	39	158	108	180	138	171	149	210	264		
	331	287	797	612	780	736	848	713	989	857	8,695	3,295 *57
CONTAGIOUS DISEASES.												
Scarlet Fever	193	192	580	604	220	226	141	180	172	172	1,346	1,334
Whooping-cough	7	11	45	48	105	103	24	54	88	49	219	265
Measles	6	4	31	37	47	48	38	41	47	38	182	168
Small-pox	21	23	282	333	105	108	56	55	95	112	559	641
	227	230	941	1,022	487	495	259	290	352	371	2,266	2,4.8 *106
FEEBLE VITALITY.												
Still-born	208	139	863	266	414	297	492	311	462	326	1,880	1,239
Debility	87	85	244	181	225	181	225	203	236	155	1,017	835
Malformation	24	12	14	3	13	7	18	16	12	7	57	33
Scrofula			39	20	39	27	27	18	15	23	144	100
Mara-mus	176	127	267	228	328	264	274	277	280	283	1,325	1,179
Teething	13	11	13	17	23	19	18	19	16	15	83	80
Cyanosis	13	12	38	25	25	15	27	19	23	24	126	95
	521	356	979	740	1,067	909	1,021	863	1,044	863	4,632	3,661 *59
Deaths from all other Causes	346	250	709	543	710	501	860	629	1,254	879	3,879	2,801 *72
Totals	2,114	1,780 *84	4,830	4,217 *88	4,246	3,772 *88	4,494	3,956 *88	5,060	4,328 *86	20,734	18,052 *87

* The small figures under the footings of females show the proportion per 100 of males.

While it has been deemed desirable to include the foregoing table, it is available as evidence of the increased mortality of boys under two years of age only in the few diseases which are confined to the period of early infancy, as cholera infantum, marasmus, cyanosis, etc. Statistics show that after the second year up to the tenth the deaths of the two sexes are about equal in proportion. From the tenth to the twentieth years the excess of deaths is a trifle greater among females. This must not be forgotten in considering the preceding table. For instance, of the 207 deaths of infants by small-pox in 1864, only 68 were of those under two years of age. Of the 244 deaths by scarlet fever only 76 were of infants under two years. Of the 140 deaths from dysentery only 78 were of children under two years, while of the 641 deaths by cholera infantum 613 were of children under two years.

But enough can readily be gathered from the table to confirm the conclusion that there is a much greater tendency to fatal diseases in the early period of existence of boys than of girls. Take first the still-born, in proportion to each 100 boys only 71 girls, then those deficient in vital stamina (marasmus, debility, etc.), to each

100 boys only 86 girls; followed by the brain diseases, to each 100 boys only 87 girls. Diseases of the respiratory organs and those of the contagious class seem by the record to be more fatal to girls than to boys, but it must not be forgotten that the table includes all under 20 years of age, and the records showed that much the larger number of cases of these classes of disease were of infants over two years of age; particularly was this so of consumption, where much the largest number of deaths was of those between 15 and 20 years—a period of life known to be especially fruitful of deaths from this complaint among young women. Besides it must be remembered that the male children have already been twice decimated by the other diseases mentioned; that while at birth they exceed the girls by five to seven per cent., at ten years of age the positions are more than reversed, the girls exceeding them by 10 to 15 per cent., so that later when the proportions of deaths are equal, the deaths of girls are indicated by increased numbers.

The tendency of whooping-cough to be more fatal to girls than to boys—as in New York City —is noteworthy; also the tendency of jaundice to being more fatal to boys, the cases reported being to each 100 boys only 66 girls.

Incidentally, an inference may be drawn from the comparative tendencies of the different organs of the human body to disease, that the more highly organized matter of the brain—that which in its intensity and power makes the animal man a human being—is more liable to give way in disease than the more brute functions of respiration and nutrition, both of which hold out longer. And that it is not probably true, as often held, that when disease fastens upon the over-excitable brain of a child, that it is because the brain is too powerful for the rest of the body, but rather that it is too impressible, and lacks consistence and strength to sustain its work in the economy of the whole body.

The main purpose, however, in the presentation of the vital and mortuary statistics in this chapter, is to give the cause for the larger number of women in the communities ; to show that in certain conditions of life there is an increased proportion of girls born, and in the same conditions, and probably from the same originating cause, a marked increase in the proportion of boys dying in early infancy, both combining to produce the result.

The first step toward overcoming any evil in the community, is to get a clear understanding of its cause. So long as we look to such secondary and transient influences as emigration to account for the deficient proportion of men, so long will we fail of attempting even to cure the evil.

If the reader is not yet fully satisfied to accept my conclusions as above, I ask him to bear with a few more dry figures in summing up the case. For this I revert to the Massachusetts statistics, the only case where the classification of details is sufficient to make the calculations proposed.

If we take the thirteen towns mentioned where the birth of girls was greatest in proportion to the boys, and deduct the numbers of both sexes that will have died by the end of two years, taking only the general proportions of the State for this, there are left 1,045 girls to each 1,000 boys. If we take the six towns mentioned in the list where the proportion of births of girls was still greater, and make only the same rate of deductions for deaths of each, there are left at the end of two years 1,140 girls to 1,000 boys. The real condition at the end of two years in such localities is worse than this, inasmuch as where so large a proportion of girls is born the deaths of

boys in the two years are greater than the general
average indicates, and really in these localities
there would be 100 to 150 boys less than the figures named. So that the condition may not be
exaggerated, take, as the fair mean at the end of
two years, 1,100 girls to each 1,000 boys. Now
it is in accordance with our general observation
and with statistics everywhere, that there is ordinarily a difference in the ages of the husband
and wife of at least five years. And statistics
further show that the natural increase of the
population by births is about three per cent.
The boys born in 1854 will be in 1879 twenty-five
years of age, and marriageable with the girls
born in 1859. But the number of these have increased in five years × three = 15 per cent., and
will stand when twenty years old in 1879, rela-
tively to the men of twenty-five years, as 1,265
to each 1,000 men. Then we must consider the
risks of deaths among the men for this extra
five years, which at this period of life is about
one per cent. per annum, so that 50 of the men
will have died out of the 1,000, reducing their
number to 950 ; making the relative proportions
at the marriageable age, 1,000 men to 1,335 wom-
en. This relative proportion will be reduced

when the average difference in the ages of the husband and wife is less than five years, and of course increased when the difference in years is greater. It needs a very trifling allowance for the influences of emigration to warrant the conclusion that, if not now, it will be true in a very few years, that one out of every three young women in these Eastern localities will have to remain unmarried because there is no man existing, in her near neighborhood at least, to mate with her. A conclusion most melancholy and lamentable for the society in which this is, if not for the young women themselves.

The estimate in the previous paragraph of the proportion between the men of twenty-five and the women of twenty that would exist in 1879 in these thirteen towns in the State of Massachusetts, was made over ten years ago. It is well supported by the U. S. Census of 1880, which gives the number of men in the whole State 25 to 30 years old as 75,212, and the number of women 20 to 25 years old as 99,589, a proportion of 1,000 males to 1,324 females. The estimate for the thirteen towns started on the basis of 975 girls born to each 1,000 boys, while at the same time there were in the State at large

only 940 girls to each 1,000 boys. It is a fair
presumption, therefore, that if a separate enrol-
ment were had in the Census of 1880 of these
thirteen towns, they would show at least this
proportional difference, making the record for
them alone 1,373 women of 20 to 25 years, to each
1,000 men of 25 to 30 years.

Emigration from these thirteen towns has not
been taken into account as a factor in the problem. There are no statistics of this obtainable,
and the reader must make his own allowance.
Observation, however, leads to the belief that in
places where women are so greatly superabundant it is difficult for them to obtain remunera-
tive employment, and they change to other localities quite as freely as the men.

The thoughtful reader will readily perceive
that this condition of affairs is now comparatively in its infancy. It does not require that
the mothers of many families in a city or town
are under the influences inducing especially the
production of female offspring, to bring about
all the disproportion that now exists.

Owing to the very limited reports of births in
New York City twenty years ago, it is not possible to make any similar calculations from the

mortuary table for that city of the proportional numbers of each sex that would be living now. Personal observations and investigations, too vague and indefinite as yet to present here, induce the belief that among the native-born mothers, the results in this respect will approach very nearly to those found in the sixteen effeminate towns in Massachusetts.

In Philadelphia, owing to the deficient classification before mentioned, classes of mothers can not be selected, nor can special localities in the city, to show and compare varied results in the proportions of the sexes dying in early infancy, if such variations exist. The mean as given in the table shows a larger proportion of male births than either Massachusetts or New York City, viz.: 1,000 boys to 911 girls. The deaths of all under 20 years for the four and a half years given are 20,374 boys, 18,053 girls. Of these 14,800 were over two years of age, and if these be taken equally from both sexes, which is about in accordance with the New York City and Massachusetts statistics, it leaves the death-roll of those under two years, 13,334 boys to 10,653 girls. This would leave the proportion at the end of two years from birth of 1,000 boys to 975 girls. Taking these, as on a previous page, marriageable—the men at 25, the women at 20—and adding as before for the natural increase in five years of the women, say fifteen per cent., and deducting for the deaths among the men for the five years, say five per cent., as before, and at these marriageable ages of 25 and 20 years the proportion will be to each 1,000 men 1,180 women.

The statistics, therefore, tend to show that while Philadelphia has reached a point in the enervation of the mothers where the male children fall victims to disease as readily as in New York City, or in the Massachusetts towns, it has not yet reached the second stage, where the proportion of male births is reduced, as in the other places named.

The foregoing estimate of the proportion between the women of 20 and the men of 25 that would be living in Philadelphia in 1879 was made, like that some four pages back for Massachusetts, over ten years ago. Its correctness can not be verified by the census of 1880 as was that, the published record not showing the numbers of these ages except for the whole State. For the city alone (Philadelphia County) that census gives of all ages 405,975 males to 441,195 females, a proportion of 1,086 females to each 1,000 males.

A paragraph is going the rounds of the newspapers this summer (1884) stating that the last census shows there are 30,000 more marriageable women in that city than men. I am not able to verify this, but counting all persons over 17 as marriageable, it will give about 1,115 women to each 1,000 men. I incline, however, to believe that a separate classification of the ages named would show even a larger proportion of women than my estimate.[1]

---

1 See Appendix, Note B., page 152.

CHAPTER III.

BEFORE fully developing the author's theory on this subject, it may be well to give some consideration to the commonly received opinion that the decision of this matter of sex rests with that Divine Providence who first created and now governs the universe, and who arbitrarily causes the embryo man to take on the male or female condition according to the dictates of His own sovereign will. As remarked in the introductory, to many this conclusion is the end of all investigation; and if we could all accept the belief it should end further investigation for all of us. I will endeavor, however, reverently to show that this must be an erroneous conclusion, and whether the theory presented in this book be the true one or not, feel well assured that the reader will not close it without at least coming to the belief that there is some natural law governing this

(44)

phenomenon, as there seems to be in most others in nature.

Too often this attribution of mysterious events in nature to a Divine Providence as the direct agent, is but the ignorant conclusion of a mind incapable of further research. In the effort to discover the occult law which governs some mysterious action in nature, he traces up the various steps of intermediate causes till he finds himself baffled, and then falls back on the direct intervention of a Divine Providence for what is yet wanting to explain the mystery. Now, such conclusions are often very depreciative of the power of the great Creator of the universe. This is evidently His Sabbath of rest. The whole creation, so far as science has revealed it to us, moves on harmoniously by virtue of certain natural laws which its Creator has put in force; under His supervision as the Great Engineer. And to conclude, that He who created the sun, stars, and revolving planets, and keeps them all moving harmoniously in space by the simple law of gravity; who implanted in each animal and plant the ability to reproduce itself *ad infinitum;* who gave to man reason, and to brutes instinct by which they when once in existence continue

to appropriate from day to day the food they require, which going through another wonderful
process of digestion nourishes and keeps them
in being year by year, and who had the wisdom
to arrange for the continuance in being of His
organic creation in a thousand separate and special ways for their comfort and happiness, had
not the power or wisdom to create a general law
by which the two sexes should be appropriately
proportioned, but must perforce exert a special
and direct act or energy at the conception of
each of the myriads of new beings in man, beast,
bird, reptile, and fish, to make it male or female,
is surely very depreciating to the Divine attri-
butes of knowledge and power. It is as though
the inventor of a clock striking the hours, had
not intelligence enough to put a cam on the
wheel carrying the minute-hand to set the striking machinery in motion, but must stand by and
pull a wire himself at the right moment to set
these in operation.

True, man is justified in thinking himself of
more importance to his Creator than are the
beasts of the field and the birds of the air, and
hence in a measure excusable for the belief that
in the offspring of his own species, when an im-

mortal soul is brought into existence, the condition of sex may be a special intervening act of the Supreme Being. But though it may be humiliating to our pride, we can not be blind to the fact that we are subject to the same natural laws in our existence as govern the inferior orders. And it is more reasonable, more reverent to believe that there is some great general law which determines the sex, not only in the plant, the insect, the reptile, fish, bird, and beast, but in the human species also. Accepting this as so, and that it is a universal law throughout all organic nature, the subject is followed up by a reference to some of the limitations there must be to its action, and an endeavor to show how and when it acts, and when not. It is the theory of some that a female is simply an undeveloped male ; that the female organs are the same as the male, only reverted, or not fully developed ; and, influenced by this resemblance, are led to believe that there is something in the nourishment of the embryo which induces its retardation into a female, or its fuller development into a male. This is simply a phase or part of the development theory, and finds some confirmation in a fact noticed by breeders of domestic cattle, that the male frequently takes some days longer for its gestation than the female.

But if we adopt one general law governing this, we see in the propagation by an egg, that the hypothesis of a special fœtal nourishment as the cause for sex must be erroneous. The embryo having its supply of food bound up in the shell with itself, must be entirely independent of the mother's condition. Even if in most of the species produced by the egg it may remain in the ovary of the mother long enough after impregnation to have some influence of this kind produced on it, we can not disregard the condition of fishes, where the ova are entirely separated from the mother fish, and beyond her influence before they are impregnated. In this order, as is well known in the art of pisciculture, it is not necessary that the male and female parents of the young fry are even brought into contact. The eggs of the female being stripped from her into a tank of water, and the milt or semen of the male in like manner stripped from him and mixed in with the eggs they become impregnated, the result being the hatching-out subsequently of both male and female fish. In like manner in the vegetable world, it needs only the bringing together of the products of the male and female flowers to produce a new plant with all the characteristics of the parent plants.

We must, therefore, look for the operation of this general law either before or at the moment of conception or impregnation, and not afterward.

This universal law must be something different and more potent than mere animal desire or lust, that thus brings together the two sexes of man, beast, bird, fish, and plant. For they presuppose a brain and nervous organization, and we can scarcely conceive these to exist in the milt and ova of fish, and certainly they do not exist in the pollen and ovule of plants.

What this energy probably is will be shown in a subsequent chapter.

3

CHAPTER IV.

STAGES OF THE INVESTIGATION.

THAT the reader may fully understand the various stages of the investigation that finally resulted in the discovery of the natural law influencing the production of sex, and be able by making independent observations of his own to judge of the reasonableness of the conclusion, there will be given in this chapter a few of the observations noted by the writer, as the foundation or frame to the work. But few of the many noticed are given, enough only to show their scope and direction, and these will be followed with what may be called general conclusions, as founded on or deduced from the observations.

As will be seen, the observations were specially directed to what seemed abnormal and irregular in the production from one pair of an unusual number of males, or an unusual number of females, and especially cases where the irregularity was hereditary. These are numbered in order, so they may be readily referred to later.

1. A married couple whose ancestry were of good healthy and robust stock, had children— first a son, then a daughter, then another son, followed by seven daughters in succession. The father was of a family of two sons and one daughter, the mother from one with five sons and two daughters. The elder son of this pair married at 40 a woman of 19, and had four children—all daughters. The eldest daughter married at 24 a man of her own age, had first one son, then five daughters. The second son married at 26 a woman of 23, had three sons and three daughters somewhat intermixed in the order of birth. The second daughter married at 20 a man of 28, had no children. The third daughter married at 24 a man of 26, had only two children, daughters, she dying soon after the birth of the last. The fourth daughter married at 21 a man of 27, had in succession five daughters, then two sons, then another daughter, followed by another son, the eldest daughter and all three sons dying in early infancy. The fifth daughter married at 20 a man of 24, had only two children, daughters, both of which died at birth, or shortly after,

the mother also dying a few days after the last infant. The sixth daughter married at 20 a man of 25, had three sons, six daughters alternating somewhat in order. The seventh and eighth daughters died young unmarried, one by an accident, the other by a contagious disease. Only a few of the third generation of this family are married, but there is quite an undue proportion of girls born from those that are.

2. Another married couple of good healthy parentage had two sons and two daughters, alternating in the order of their birth. The two sons, marrying, had born to them eight daughters and three sons. The two daughters, marrying, had born to them seven sons and one daughter only.

3. Another married couple had first born to them two daughters, then five sons in succession. The father had a severe sickness after the birth of the second daughter, from the more serious effects of which he recovered, but was somewhat of an invalid thereafter.

4. Two sons, the only children of one family, married two sisters, the only children of another family. The antecedents of both families not known, but the several children born in both families were girls only.

5. An acquaintance in Pennsylvania, whose wife had borne him several children, mostly girls, was elected a member of the House of Representatives. He spent the winter in Harrisburg; this being before the time when railroads ran to every section of the State, travel home was to him both tedious and expensive. In exactly nine months after his return home in the spring, his wife had a son. After spending the second winter in Harrisburg he was elected to Congress, and the third winter was spent in Washington. Again, promptly in nine months after his return home in the spring, his wife had another boy. The two events were thus so peculiarly marked that among his intimate friends the boys were dubbed the " Representative " and the " Congressman."

6. A neighbor of the author, a milkman, had a herd of some twenty or more milch-cows, for whose service he kept a bull, and somewhat, also, to serve his neighbors' cows. He said it was a very unusual thing for a heifer calf to be born among his herd, sometimes not one the whole season ; while among the cows of the neighbors served by the same bull, the calves were no more frequently male than female.

7. Two young sows of the same litter belonging to the author—as near alike as two peas—being in heat, were driven one morning about half a mile to a neighboring farm where a boar was kept. One of them was served and driven home. But as the boar showed no inclination for another immediate union, the other sow was left on the place with him to be served and sent home later. The litter of the first one served was, six female, two male; of the second one, seven male, two female.

These show the character of the observations. It is not necessary to multiply them, for similar cases will, no doubt, come to the mind of the reader on reflection among his own experiences. After numerous observations of this kind had been made where opportunities were had to study the surroundings and influences that might have operated to produce these irregularities in the production of sex, it seemed possible with the aid of the statistics given herein in chapter first, to collate and condense them all into some general conclusions, as follows:

a. Robust, healthy, and apparently lusty wives more frequently have male than female children, particularly so when their husbands are of medium or inferior vigor, and reversely:

b. Delicate, weak women, who indicate from their appearance but little sexual ardor, more frequently have female children, especially when their husbands show indications of greater sexual vigor.

c. Women who have these characteristics in a medium degree, which may fairly represent the great majority of wives, and whose husbands are also of fair average vigor, if they continue having children regularly about every two years, will have more girls than boys.

d. Wives who have been brought up religiously, and when young girls become devoted church-members, have usually a larger proportion of girls than boys. This has been as yet mainly noticed in village communities.

e. The wives of the farming population have more boys than girls, and reversely :

f. The wives of a city, town, or sometimes even a village population, give birth to more girls than boys.

g. Illegitimate children born throughout the country are in very large proportion boys, as much, or more, than 3 to 1.

h. Illegitimate children born in cities, though oftener boys than girls, are not so to near the extent they are in the country.

i. The children of a woman 18 to 22, who is married to a man 35 to 40, are in larger proportion girls.

j. The children of a woman of 25 to 35, whose husband is 5 to 10 years younger than she is, will have a larger proportion of boys.

k. In a family of brothers and sisters, if the sisters when married have a preponderance of girls, the brothers when married will have a preponderance of boys, and reversely:

l. If the sisters when married have mostly boys, the brothers when married will have mostly girls.

m. The begetting of girls requires so much of sexual vigor in the father, that a saying has become common among the rural population, "that any boy can beget a boy, but it takes a man to beget a girl."

CHAPTER V.

AFTER much casting about to discover some physical law or laws that would, if applied, cover all the observations made and general conclusions drawn from them, the following was eventually settled upon : That at the generation of male offspring the mother must be in a higher degree of sexual excitement than the father. And reversely, at the generation of female offspring, the father must be in a higher state of such excitement than the mother. A remark made to the writer by a countryman of his acquaintance, with whom he was conversing on the subject, somewhat accidentally led to this : it was that he could always tell when his wife was conceiving a boy, for she did all the work ; while if it was a girl, he had to do all the work. While this may be regarded as an exaggerated remark, and such one-sided activity seldom required, it

3*   (57)

does in a forcible way reveal the prominent idea, and worth remembering by those wishing to test the theory.

I am fully aware that this is not the prevalent view, and the one which perhaps seems more natural, that each parent impresses his, or her, own sex on the offspring, the strongest one governing the influence on the embryo. But this view has not been held as the result of any investigation, or because it seems to meet the necessary requirements, especially of the abnormal or irregular cases, where there are many more of one sex than of the other from one pair. It passes simply because it seems natural and reasonable, and is supposed to account for the varying sex in a family where the husband and wife are about on an equality of sexual vigor, and the children fairly divided between male and female. But even here it is not the true principle, for it does not cover the oft-repeated observations of particular instances of irregular proportions of the sexes from one pair, while the reverse does. This it is purposed now to show by taking up the specially observed cases mentioned in the preceding chapter, and the general conclusions *seriatim*, and indicating from the circumstances surrounding each, how the hypothesis fits them. Each observation and conclusion is referred to by its original number and letter, so they need not be repeated in detail.

Observation 1. As mentioned, the husband and wife were both of good healthy stock. The husband had a brother and sister only, the wife four brothers and one sister, none of them showing any lack of physical vigor, and there was nothing to show any inherited tendency to such a large proportion of daughters as they had. The first four children born being alternately a son and a daughter repeated, indicated that the parents were well mated as regards sexual vigor. But about this period of her married life the wife became what may fairly be called by its common name, complaining, and from year to year grew worse, though having a girl about every two years. From and after the birth of her last daughter she was a helpless invalid, dying when her youngest child was about twelve years old. During all this time of invalidism she necessarily ranked much below her husband in sexual vigor, and, as we shall see, somewhat entailed upon her later offspring especially, a debility of organization that resulted in their

cases in the production of an excess of girls.
The elder son, marrying at the mature age of 40,
when in his full sexual vigor, a young and
scarcely mature woman, was naturally much her
superior in sexual vigor ; their four children succeeded each other two years, or about that,
apart, were girls, partly because of the superior
strength of the father, and partly because the
mother by her continuous child-bearing and lactation had no opportunity of fully recuperating
her strength.

The first daughter born before her mother had
fallen into this " complaining " state, inherited a
good degree of robust health, but being married
to a man of remarkable vigor, had first a son,
and later children so regularly every two years,
that her strength was not re-established after
lactation before she again conceived, so that her
progeny were later all girls.   The second son, a
fairly vigorous man, married a woman fully as
strong and vigorous as himself, resulting in offspring equally divided as to sex.

All the rest of the daughters who had children
seemed to have inherited a weak sexual organization from their mother, and produced almost
altogether girls, except the youngest one.   When

she was yet but a child, owing to the poor health
of the mother, the family moved to the country,
where she, as a growing girl, enjoyed unusual
advantages of much outdoor life, while her older sisters were confined either at boarding-school,
or in the care and attendance on the invalid
mother.

Obs. 2. This was a case where the vitality of the
family was of the highest, and the husband and
wife well balanced as to sexual vigor, as shown
by an equal number of boys and girls. But the
sons, in marrying, proved to be more vigorous
than the average, and their wives only moder-
ately so. The consequence was, the birth of
daughters mainly to both of them ; while the
daughters, on the other hand, possessing good
vigorous constitutions beyond the average, found
in their husbands men of less vigor, with the result that their children were mainly sons.

Obs. 3. This was a case where naturally the
husband was at first of more vigor than the wife,
so they had first two daughters. After the husband's sickness and invalidism his wife was the
more vigorous of the two, and they then had
only boys.

Obs. 4. Here the two daughters had inherited a

delicacy of bodily organization which placed them below the average, and being married to two only sons who had inherited a good share of robustness from a mother strong enough to have sons only, the offspring on the hypothesis mentioned would likely be, as they were, girls only.   ·

Obs. 5. This case shows to the reader how a tendency to have girls born in a family may be overcome in a natural and proper way. That is by a total separation of the husband from the wife's bed for a time, longer or shorter, to enable her to fully recuperate her strength, and to get up a strong, healthy, natural desire for a reunion with her husband. In this case not only was the separation a fitting one for this purpose, but the coming together was under favorable circumstances for the wife's conception of a son. The separation had been long enough to kindle in the wife a strong desire for the embrace of her husband. Ordinarily his would have been proportionally stronger, and the chance of a male conception less probable ; but this was away back, before the time when railroads ran to every little town, and the husband reached home after some three days' travel by stage-coach and wagon. It is not necessary to suppose that he had during the separation from his wife illegitimately satisfied his desires, and was, therefore, less sexually excitable than his wife. Fatigued by his long ride, he would naturally feel more like resting and sleeping on his retirement to bed, than indulging very effectively his wife's ardent desire,—a condition on both sides, according to the hypothesis, for the conception of male offspring.

Obs. 6. These facts all accord with the hypothesis. The milkman was eminently shrewd and practical in the management of his herd. It was always composed of cows in their prime ; they were kept in the best possible condition of flesh, and with very little stall-feeding were always fit for the butcher. He did not think it profitable to raise the few heifer calves his herd produced, but as any of his cows got past their prime they were sold off to the butcher, and their place supplied with others younger but still at their maturity, these giving the best results in the production of milk. Consequently his cows were always at the best and most vigorous period of their existence, neither very young nor very old. For the same economic reason he raised every year or two a bull-calf, which, as soon as old enough, served his herd of cows, but which so soon as he had fairly his growth, and while still young, was castrated and sold to the butcher, his place being supplied by a growing younger one from time to time. The bull did not run with the cows, but when one of them was in heat she was turned into the small enclosure where he pastured, and, as he had so many cows to serve, only a single intercourse was allowed between them. The conditions were such as to give each cow separately a much higher degree of sexual excitement at the intercourse than the bull possessed. In the first place, his ardor was often weak from having so large a herd to serve ; and in the second place, there was, so to say, no preparation or excitation of the bull prior to the connection with each cow, as there would have been had he run with the herd; while with each cow there was somewhat of this excitation from other cows of the herd, before her state of heat would be noticed by the attendants. Add to this the condition of the bull's youth and more tender muscle, as compared with the run of the cows, and the circumstances all were favorable to the engendering of

male calves. When the neighbors' cows were
sent in to the bull the conditions were often
quite different. These were often in a lower
condition of keep, had been led or driven such
distances and under such circumstances, as sometimes to greatly fatigue them, and were often
young heifers, so that at the connection of the
bull with any one, he was likely to be in as high
a state of sexual excitement as the cow was, and
there was an equal chance of as many female as
male calves resulting from the intercourse with
these neighbors' cows.

Obs. 7. This case is explainable on the hypothesis as follows : The sows in being driven
to the boar by a way somewhat tortuous, with
branchings-off from the road, required some rac-
ing to keep them in the right track, from the
natural contrariness of the animal that always
wants to go the way different from that the
driver wishes. So they arrived at the home of
the boar somewhat tired out, and without opportunity to rest an immediate union was had between one of them and the boar while he was
fresh and not fatigued. According to the hypothesis, under these conditions female pigs
would be more likely conceived, all other things

being equal. The second sow being left on the premises till later, had ample time to rest and regain all her natural strength. The later union between her and the boar while he had yet scarcely recovered strength after his previous encounter, naturally gave this sow the chance to be in a higher degree of sexual excitement than he, resulting in the conception of a larger proportion of male pigs in the litter.[1]

---

1 See Appendix, Note C., page 154.

CHAPTER VI.

THE GENERAL CONCLUSIONS IN CHAPTER FOUR REVIEWED IN THE LIGHT OF THE PHYSICAL LAW ENUNCIATED IN CHAPTER FIVE.

a. Not alone the statistics given in chapter second, but observation shows that a larger proportion of boys are born in country places than in towns. Among the rural or farming population the labor of the wife is not so fatiguing as that of the husband. To a healthy woman it is not greater than is fairly needed for exercise to keep the body in good physical condition. And when the hour of rest comes at which the intercourse usually occurs resulting in a conception, the wife is more likely to be active and vigorous than the husband, who, from following the plow all day, or other arduous farm labor, is tired and exhausted. And naturally the wife takes on more readily the higher degree of sexual ardor, and most frequently has male offspring.

b. A different condition exists in towns; the

wives lead less active lives, and do not have exercise enough, nor fresh, wholesome air enough
to give firmness and strength to their muscle:
while the husbands in great measure have
enough of these to give them a fair degree of
muscular strength without so much as to produce bodily exhaustion. Consequently on retirement at night, as a general thing, they will be
more vigorously excited sexually than the wives,
resulting in a larger proportion of female conceptions. [See the statistics for the condition of
towns in Massachusetts.]

c. This condition is accounted for under the
physical law as follows : Where wives have children in regular marital intercourse every two
years or thereabout, the periodicity is usually
from their nursing the previous child for a year
or more after its birth, and not menstruating
during that period. That this function ceases
during lactation shows that they are not specially robust. Some fruit-trees, especially the apple,
when in an impoverished soil can not nourish the
fruit, and form new fruit-buds for the next year's
supply at the same time. Hence the tree bears
a crop every other year only. The most of our
fruits would come in this way, were it not that

in some the fruit matures and drops off so early
in the year, that there is yet time afterward in
the autumn for the growth of the fruit-buds.

When at the end of a year's nursing the wife
finds the increased demands of the child for sus-
tenance is taxing too greatly her strength it is
weaned. Then menstruation soon commences, and
she again becomes pregnant while her strength is
not yet recovered. There is scarcely a chance that
her sexual passion will be naturally aroused at
the time of conception again, when it takes place
thus promptly following the cessation of nursing. Even in the case of stronger wives who
while partly feeding their child artificially, yet
continue also a partial nursing, during which
menstruation commences, the existence of both
functions reduces their sexual power to a feebler
condition, so that they too will conceive girls
where, freed for a time from the first of these
drains on their physical powers, they would ordinarily conceive only boys.

The same train of circumstances follows year
after year, or rather every two years, and wives
otherwise robust enough to have only male offspring have a succession of girls. This succession
is sometimes broken by accidental circumstances,

that separate the husband from the marriage-bed
for two or three months at about the time when
conception would ordinarily occur, and this being thus delayed, the wife has an opportunity to
recover her full vigor.[1]

d. This will, no doubt, be regarded as a strange
conclusion, but if any of my readers will take the
trouble to investigate the subject they will find
it a true one. The fact is explainable under
the physical law as resulting from two primary
causes. First, a girl trained up religiously is
taught that even thoughts on the subject of procreation are wrong, and are to be suppressed.
She grows up with an innate modesty and pudic-
ity, often arriving at a state of puberty without
a sensual thought or inclination toward the opposite sex. She even feels an inward shame when
nature develops the sexual desire in her, lest
some unconscious word or act should betray to
others its existence. And when she is married
this modesty has become so ingrained that it
controls her actions in the conjugal embraces of
her husband. She prefers to have him think
that she submits to these to please him, and as a
matter of duty, rather than from any special desire on her part.

---

1 See Appendix, Note D., page 155.

So it happens that these embraces are rarely sought by her, or her desire for them indicated, and they occur only at the solicitation of the husband when his desires are ardent, and when often her own are not. At such connections the chances are very few in proportion that she will be in the higher condition of sexual excitement and conceive a son.

If later on in the married life of such wives, they somewhat lose this prudery, they are likely to be found in the condition mentioned in conclusion (c).

But, secondly, there is another cause for the condition of such wives as leads to this conclusion (d) lying much deeper. Young women of poor health and feeble bodily frames more frequently and readily become church-members than do their more robust sisters. This weaker condition naturally leads to thoughts on the probability of an early death, and they hasten to make preparation for the life beyond the grave by joining the church ; while their sisters, with bounding, vigorous health and active desires, feel that death is yet a long way off. So the solemn exhortation of the preacher to prepare for it, which moves the feebler ones to re-

pentance, and to church-membership, passes by
these stronger ones unheeded. The sunshine and
glamor of life is yet all before them, and thoughts
of the grave and the hereafter are for them at
this period of their lives but the passing fleecy
clouds in the sky of their bright June days, that
shadow them but for the moment.

Readers may regard this conclusion (*d*) as a
sad one, but it is a true one, as any of them
will find who may have access to the records
of infant baptism in any of our village churches
where our native-born women are largely members; the female infants baptized will be two, or
more even, to one male infant. Of course, this
idea of the tendency of those in feeble health to
more readily devote themselves to a religious
life is not new. Holmes in his " Professor at
the Breakfast Table " discourses upon it, and
upon the tendency, on the other hand, of the
young in robust health to be mischievously
wicked; closing his discourse with the remark
that—" in the sensibility and sanctity which
often accompany premature decay, I see one of
the most beautiful instances of the principle of
compensation that marks the Divine benevo-
lence." So Longfellow in one of the most beau-

tiful of his short poems, "Footsteps of Angels," connects delicacy of bodily frame with saintliness in—

> " They the holy ones and weakly,
>   Who the cross of suffering bore,
> Folded their pale hands so meekly,
>   Spoke with us on earth no more."

There is a like conclusion in the old anecdote of the little boy who from reading his Sunday-school story books so coupled goodness with an early death, that when exhorted by his mother to be a good boy, replied that he "did not want to be good, for the good little children all died and went to heaven."[1]

Admitting this second view of the greater tendency of weak and feeble young women becoming church-members, we have the reason under the physical law stated, why, when they become wives to men of average physical stamina, there should be many more girls than boys born of them.

e. f. Conclusions (*e*) and (*f*) are really included in those of (*a*) and (*b*)—the former being general and these special ; consequently the explanatory remarks connected with (*a*) and (*b*) cover also (*e*) and (*f*) so far as showing how the general law of

---

[1] See Appendix, Note E., page 158.

the conception of sex is applicable. They are
kept separated because the statistics more especially and directly show the conclusions (e)
and (f) to be true, by the details of the towns
given.

Other general conclusions of similar character
to (e) and (f) were arrived at from neighborhood
observations—as that a larger proportion of girls
were born in well-to-do families where the wives
being entirely freed from all active bodily labor
devote their time to quiet sedentary occupa-
tions, as embroidering, crocheting, and the like,
and become weak and languid in muscular fibre,
than there is among poorer families where the
wives do their own household work, and by the
exercise therein retain more physical strength;
but all such conclusions were finally embraced
in those given.

g. h. These are general conclusions formed
from personal observation as to (g), for no statistics were found of illegitimate births in the
country places. Necessarily the horizon of any
one person's observations of this kind is extreme-
ly restricted. The few instances of such births
occurring within the writer's notice being at least
three boys to one girl, justified the conclusion,

though it was not justified of illegitimate births recorded in the cities.

The reason for this doubtless is that illegitimate births throughout the country are more frequently the result of the seduction of the woman by her pseudo lover. Naturally a transgression of this character would not occur except when the woman was in so highly excited a state as to be largely regardless of the consequences. The opportunities would not be frequent between the pair for their illicit intercourse, and whenever they did occur the chances would be in favor of the woman's being in such condition that, under the physical law given, she would more readily conceive male children.

In cities, while this condition exists, there are two other classes of women who have illegitimate children under different influences. One class, the prostitutes, whose frequent indulgence of the sexual passion naturally hinders it from ever rising to a great height; while their temporary paramour is likely to be more vigorous than the ordinary husband is in the marital intercourse. Conceptions occurring under these circumstances would not probably vary much in the proportions of the sex of the offspring from

those occurring among the married people of the
same place.   The other class having illegitimate
children are the mistresses, who occupy a similar
position to that a wife does in regard to their
chance of the conception of either sex.   The disposition of men of mature age to select very
youthful women for mistresses would indeed
result in more frequent births of girls than of
boys, as mentioned in general conclusion (*i*).

An article in the New York *Mail and Ex-
press* July 25, 1884, on " Homeless Waifs," says :
" Three-fourths of the abandoned babes that are
picked up in the streets are boys.   Of the eight-
een foundlings brought to Matron Webb's Nur-
sery " (Police Headquarters) " in the first eight-
een days of July fourteen were boys, and of the
twenty-two in June, thirteen."   It is a fair presumption that these abandoned babes are mostly
illegitimate.

i. It is a somewhat prevalent idea that the sexual ardor is much stronger in youth than in mature age.   This is, however, an error.   That the
passion is more excitable in the young and less
under control, is no doubt true.   In youth the
tissues are softer and more sensitive to all impressions ; the sensation of pain is felt more

acutely, so also the sensation of pleasure. But this sensibility to the impression is a different thing from the strength of the impression, and the power that gives it. The one presupposes softness of muscle and tenderness of nerve, producing an excitation at less force of power. The other hardness of muscle, slow to excitement, and, consequently, the exertion of more sexual force to arouse it. It is this last that gives the superior power in the sexual embrace. All other things being equal, therefore in the union of a husband of 35 or 40 years to a wife of 18 to 22 years, it would be very seldom that the wife would be in such a superior sexual ardor to her husband, at the time of the conjugal embrace, as to conceive a male child. She might, and probably would be, more frequently in the condition of amatory desire than he, but rarely rising to the same height; just as a child will be more frequently and more excitably hungry than a man, yet without the man's ability to eat or digest the quantity of food.

j. Marriages do not often occur in this, the reverse condition of the preceding ; but they occasionally do, especially among the Irish population, where the union of a man of 21 to 24 to a woman of 30 to 32 is not uncommon. In the few cases I have noticed of this character, the children have been in large proportion boys.

k. The fact that there is a fair proportion of brothers and sisters in a family, indicates a fair equality of sexual vigor in the parents, but is no proof that they in this vigor are above or below the average. To whatever degree it existed it would be inherited by the children. If the daughters on being married, having no inherent defect of constitution, give birth to a larger proportion of girls, the inference under the natural law given is, that as a family they are below the average of sexual vigor. The brothers, partaking of the same inferior condition, wedded to wives possessing this vigor in the average, or to them superior, degree, would have by their wives a larger proportion of boys.

l. In this, the reverse condition, if the daughters give birth to a larger proportion of sons, the inference is the family vigor is above the average. And the sons inheriting this higher vigor when married to women of average vigor, will have more girls than boys.

m. Of course this may be said at times as a sort of defence to the charge of weakness in the husband where his wife has many daughters, the popular belief being, as before mentioned, that it is the father that gives the paramount impression on the embryo when it is a son. But the author has heard it quoted with pride on the birth of a girl, when a fair proportion of the children already born in the family were boys. It will be obvious to the reader that it falls in exactly with the physical law enunciated. Among a rugged population of farmers' wives a husband had to be possessed of a higher degree of virility than the very young men, usually denominated boys, ordinarily possess, so as to overbalance the strong sexual vigor of the average wife, before he could beget a girl. If he was only a boy in this particular, the stronger sexual ardor of his wife would control the conception producing male offspring.

In closing this chapter reviewing the general conclusions, it is proper to call the reader's attention to the fact that the various influences mentioned in the conclusions often clash, producing exceptions. And these must specially be borne in mind in any efforts to account for the existing condition of disproportional numbers of offspring of either sex in his own family;

or in those of his neighbors. For instance, a brother or sister, as mentioned in (*k*) and (*l*), may from special circumstances be found much above or much below the average vigor of the rest. Or one of a vigorous family may have intermarried with a person of still more vigorous family. Or one of a weaker family with one still weaker. Or one of mature years married to a very young wife, or the reverse. All these have to be carefully noted when making such observations, otherwise the conclusions would only lead the reader astray.

CHAPTER VII.

CONDITIONS AND PERIODS FAVORABLE TO THE CONCEPTION OF MALE OFFSPRING.

In the application of the physical law controlling sex, as presented in the previous chapters, to actual practice, some consideration of the conditions under which conception ordinarily takes place in the human race, and what more favorable conditions can be obtained at this period, is very desirable.

It may safely be asserted that not once in a hundred instances does it occur premeditatedly, or with this special end in view at the time of the marital embrace, but is a result, too often an undesired one, following an embrace sought simply for gratification. This being so, it is important that these occasions should be so considered and timed, that when a conception does follow, the offspring will be of the sex desired by the parents. This it will not be difficult to do by the adoption of a few simple rules not mate-

4* (81)

rially interfering with the reasonable enjoyment of this marital intercourse.

Believing that by far the greater number of married people are desirous of having a larger proportion of sons born to them than is at present current, the rules inculcated are such as tend to the conception of male offspring.

The first and great rule is never to allow of the sexual embrace except at the wife's earnest and ardent desire. This must not be a desire merely to please her husband, which is too often the impulse, but one arising from a real craving for the gratification of her passion. Should there be a remaining feeling of unsatisfied desire on her part after the consummation of the act, it would be a favorable indication that it was a male conception, if any occurred at the time, and this remaining desire should not be quenched by a repetition of the act.

As such intercourse usually takes place at night, there should be also some consideration of the best time in the night for it. This must depend on the daily occupation of the wife. If this is such during the latter part of the day and evening as to weary her, as the care of a fretful child, and she retires to bed fatigued, nervous, and unstrung, the early night is unfavorable for the conception of male offspring, even though by thinking upon the subject she may stimulate her desires to such an extent as to wish their gratification. Especially will this early hour of the night be unfavorable for male conceptions if the husband, as often happens, has ceased his daily labors and cares at the close of day, and had time for rest and recuperation before bedtime. Under such circumstances the sexual connection should be postponed till toward the morning hour, when the wife has been strengthened and refreshed by sleep. The case is different when the wife's employment late in the day and evening has been such that she retires to the marital couch not fatigued, while the husband's labors at the same period have been such as to induce a sense of fatigue. Then the hour of retiring would be appropriate for the connection, provided, of course, the wife first desired it. There is also a period in the month that the wife will do well to consider. As is known by most people, the monthly menstruation of woman is the same in general character and purpose as the period of heat in the females of the brutes. It is, however, so much more intense that the discharge is sanguineous, and the desire created more lasting. In brutes the desire is very transient, the female admitting the male, usually, only during the periodic flow or heat, while in the woman this desire often lasts the whole month through. There seems in this stronger passion of the woman a special providence to continue the existence of the race. Did she possess only the transitory desire of the brute female, her intelligence would induce her to calmly bear this fleeting demand of nature, knowing the consequences that follow its gratification, and the increase of the race would consequently languish. But as this passion in a healthy woman may last all the month through, there is a continuous demand, more or less great, for its gratification by marriage. Especially is this so in the young woman just come to maturity and at a marriageable age, when her desires, if not at their strongest, are at least more difficult to repress.

While the sexual desire incident to and following menstruation is usually spoken of as continuing through the month, it is not universally so. The writer has been told of wives who feel this passion only during the menstrual flow, and in whom it is difficult to excite it a week later. Of course with such passionless women it is difficult to conceive male offspring. Their only chance would be with husbands whose desires are as feeble as their own ; and then the probability of the production of strong, vigorous children of either sex would be very limited.

It was one of the author's secondary theories, founded on his own personal experience, that the nearer a conception occurred to the period of menstruation the greater would be the female's sexual excitement, and, therefore, the greater probability of male offspring, and that if conception was delayed a few days after menstruation the offspring would be female. But on fuller investigation this, as a universal law, had to be abandoned. There are well-authenticated cases in medical works where wives have conceived sons even twenty or more days after menstruation. There is, too, the case of the religious Jews, among whom a fair proportion of boys is born, whose sacerdotal laws forbid the intercourse of the husband and wife till full seven days have passed after menstruation. There is also the undoubted fact found in the management of bees, that if the queen is kept from the male for some days after she is willing to accept him, her progeny will be largely males.

As a fact, women vary in the continuance of this desire—some a few days only from and after menstruation, others two weeks, and still others through the full month. That a great many lose this ardency early in the month, is shown by the current belief that they will not become pregnant by an intercourse with their husbands after two full weeks from the menstrual period.

The conclusion from all this is, that every wife desiring male offspring should carefully observe her disposition in this respect for a month or two, and notice at about what period after menstruation her desires are most intense, and thus be prepared to select that period only.

Among the causes that tend to the conception of daughters none are more influential than too frequent sexual intercourse, particularly when the intercourse is at the solicitation of the husband, and the wife has no desire, except as momentarily excited by the caresses of the husband.

Many a wife's health is ruined by the nightly, or even semi-weekly, embraces of a too vigorous husband; who thinks because his wife does not object, fearing perhaps if she did he would seek the society of other women, that it is agreeable to her. Indeed, in a limited measure it may be, and yet be very injurious to her. A husband who thus sacrifices his wife's health to indulge his own desires makes—I would say—a beast of himself,—only that this would slander the beasts; for the males of these do not press their suit when the female is disinclined.

Some may object that I name the morning hour as under some circumstances more favorable for the marital intercourse that may lead to the conception of a son, because the act is one that often causes a desire for rest or repose afterward. But there is no better or more sure rule to apply when male offspring are desired, than that the sexual intercourse of the husband and wife should have been so infrequent that fatigue does not follow the act. When this continence prevails the husband will see and understand clearly the beauty and propriety of the Psalmist's comparison of the rising sun, "Like a bridegroom coming forth from his chamber and rejoicing as a strong man to run a race." Too often now the pallid and limp husband coming forth from the marital chamber after a night of exhaustive enjoy-

ment, is shorn of all his strength, and bears no
similitude to the rising sun, or to the strong
man rejoicing to run a race.

The husband should remember that his wife's
periods of high sexual desire are immutably
fixed by her menstruation at a month or about
that apart. While his are only about a week
apart—that is, if his sexual desire is completely
extinguished by a full gratification—a natural
return of desire will occur inside of a week at
furthest. It is true an artificial desire can be
stimulated on both sides by the imagination at
intermediate times. But it is this inferior condition of desire that is especially to be depre-
cated by the wife as favoring the conception of
girls.

CHAPTER VIII.

ON SOME OF THE CAUSES TENDING TO INCAPAC ITATE WOMEN FOR THE CONCEPTION OF MALE OFFSPRING.

WHEN the husband possesses an ardor in the sexual embrace so much greater than the wife that female conceptions ordinarily ensue, it is, of course, not desirable that he should be shorn of somewhat of this ardor by fasting or otherwise, as this would tend to the general weakening of the offspring; but rather measures should be taken to increase her general health and strength, and incidentally thereby her sexual vigor.

It is not exactly the province of this work to say how this is to be done, that being rather for the family physician, who, knowing all the circumstances of each individual case, can appropriately prescribe. But it is fully within the scope of the treatise to mention the character of, and most ordinary causes for, any deteriorate condition of the wife.

There are, doubtless, some wives who are perfectly healthy in all their bodily organization, who yet have but feebly developed maternal organs, and whose sexual power it would be difficult to increase. But from some observation on that point, the belief is induced that the most of wives whose sexual ardor is feeble, have some inherent taint of disease inherited or personal, which, if cured, would greatly restore to them the vigor they lack.

If a classification were made of the various degrees in which this appetite exists in wives, it would run about as follows, viz.:

1. The pristine, vigorous vitality, that will include such wives as menstruate, in common with the females of the brute creation, shortly after parturition and during lactation, and who, therefore, have the ability to conceive and bear a child every year. Women of this class are rare in our towns, but not infrequent among the rural population. Unless mated to an exceedingly lusty husband their children are largely boys; but when so mated and conception follows soon after childbirth, and before the debility usually attending this event is fully recovered from, they can have a succession of girls.

2. Those who do not menstruate till after lactation ceases, and yet are not specially enfeebled
by nursing. This condition is the best now usually found in towns. When, as is frequently the
case, they nurse their offspring for a year from
its birth, they will have children every two years.
And if conception does not occur too promptly
after weaning their child, they will oftener have
boys than girls.

3. Those who are much weakened by nursing
their child, and recover strength but slowly
afterward. Menstruation comes on after each
period of nursing while they are still weak, resulting very commonly in a succession of girls.
If an occasional boy is born, he will be of feeble
vitality, with all the softness and effeminacy of
his sisters, and very prone to succumb in the
struggle of life at an early period. Too many of
the wives in our towns are now in this reduced
condition, giving birth to the numerous weakly
girls that grow up to be mothers in their turn
with much of inherited weakness of body.

4. The lowest and most debilitated who require all their feeble vitality to maintain their
own hold on life. Some without the ability to
bear children, or if they bring a few miserable

abortions of humanity into existence, they die off in childhood. The number of this class who are wives is not great. Many of them die off on reaching womanhood, others aware of their weak condition decline to enter the married state, or are passed by in that natural selection a marrying man often makes, because they are visibly unfitted to become wives and mothers.

It will be understood that the classes above described are only types. There are always intermediate shadings and gradations between the classes, so that any given wife may at one period of her life be fairly identified with one class, and at another period by an increase or decline of health be raised or lowered to the next class.

Fortunately there is a recuperative vigor in the human body, especially in the first half of its existence, so that when we are weakened either by congenital or personal infirmities, care and attention bestowed to that end will bring an amelioration of the affliction. Even if a full restoration to the primitive health and vigor of the race is not obtained in one generation, a wife and mother may hope that by care her children, or children's children at furthest, may be exempted from the ills she suffers.

A daughter will often inherit from her mother a tendency to female disorders, just as she will inherit weakness of the lungs, or of the digestive organs, and unless such tendencies are specially cared for, mere change of residence to a more healthful locality, and the ordinary hygienic regimen for increasing the general health will not of themselves avail,—though these are essential and necessary in connection with special means for remedying the sexual difficulties.

It is natural to suppose that a woman weak in her maternal organism, is weak in all her bodily organs.   But this is by no means always so, perhaps not even enough so to establish it as a rule. Weak lungs or a weak stomach are often found coupled with a large degree of muscular strength and health in the rest of the body.   And a wife may have lungs actually diseased, yet possessed of so strong a sexual organization that she has more boys than girls.   Still as the disease of the lungs progresses we should expect to find a gradual decline of vigor and strength in every other organ of her body.   On the other hand, a wife may have a strong bodily frame, good lungs, good digestion, and yet an inherited debility— if nothing more—of the maternal organs, that inclines to the conception of girls rather than boys.

That these female complaints and weaknesses are very prevalent, especially among the well-to-do class, is now undisputed. Miss Harriet Beecher, writing on the health of women in this particular, says : "I do not know, among all my acquaintance, a woman who has a perfectly healthy maternal organization."

The most common and least noticed of these complaints, and no doubt the foundational cause of most of the others named, is immoderate or too profuse menstruation. And yet this is evidence of a weakness rather than a disease in itself. It is not too much to assert that this seriously impairs the health and strength of half the women of our country. While it is difficult to cure the complaint when once established, much can be done to alleviate it. The difficulty is that it rarely excites sufficient attention, as not one woman in ten probably knows whether in her individual case menstruation is too profuse or not. But until the wives and mothers of our land do learn something about this grave subject, our country will continue to be filled with great numbers of sickly and debilitated women, who, when wives and mothers, will produce mainly daughters, inheriting in their turn the weaknesses and complaints of their mothers.

There is a strong analogy between this function of menstruation, as exhibited in women and the females of the brute creation, and the blossoming of plants and trees in the vegetable world—a like preparatory step to fructification. Every pomologist knows that a superabundance of blossoms on his fruit trees is not only weakening to the trees, but often the indication of an inferior crop of fruit.

It may be rather fanciful to go so far for an illustration, when we really need no other than the subject herself. The remarkable change exerted in the person of a young woman when this function comes on in its regular normal manner is patent to all observers. The girl under its influence is developed into a graceful, sensuous woman. The eyes take on a new lustre, the cheeks a livelier bloom, the lips a brighter red, the neck and bust a fuller development. "Grace is in all her steps and beauty in her eye." But when this flow is in excess, how different are the appearances: the whole person exhausted and languid, the eyes sunken and lustreless, often

with dark circles around them, the cheeks wan
and wasted, lips thin and bloodless, the neck and
bust spare and angular. A few years' continuance of this excessive menstruation turns a young
woman into an old one.

As before remarked, probably not one woman
in ten, if the question was asked her, could tell
whether in her case this monthly flow was excessive or not. The other nine would suppose it
was with them as with other women ; not knowing whether the natural and healthy flow should
be a tablespoonful or half a pint, they are at a
loss to know when it is excessive. The proper
quantity is not measurable, and depends entirely
on the robustness of the woman.

An unfailing rule, however, easily understood
and remembered, is, that it is too much when it
produces weakness and exhaustion. Every woman can by this know when to apply for medical
advice, or take precautions to abate the flow. To
understand how, in the absence of medical advice,
these precautions can be taken, it is proper to
explain that extravasation of blood from the
mucous surfaces, as the interior of the nose,
mouth, bowels, and womb, occurs from two causes.
One when there is a congestion of blood in the

locality, so that the blood-vessels are distended by the inordinate supply. The blood will then, by force of pressure alone, ooze through the distended coats of the minute veins, and continue to flow till the congestion is relieved and the pressure abated. The second cause comes on to act just at this time. The relaxed coats of the minute veins are often too weak to contract against the normal pressure of the blood, and this continues to flow on as before, not because there is any congestion, but because of the weakness of the tissue. This effect is often seen in simple bleeding at the nose. A discharge of blood begun by nature to relieve the congested parts, continues on, sometimes imperilling the life of the patient, and is only stayed by the use of strong styptics topically applied, causing the membranes to contract; or sometimes by fainting, when the heart partly abates its action, and the pressure of blood in the veins almost ceases. Among the causes tending to increase, and often to originate this really dangerous and foundational complaint, tight-lacing stands pre-eminent. By this is not meant simply that of the stays or corsets ordinarily worn by women, but any and all clothing that compresses the

5

body at the waist. All dress of this character acts as a ligament, preventing the blood forced by the heart's action to the lower part of the body, through the arteries, from readily returning back through the veins. The effect is precisely the same as when the ligature is applied to the arm by the surgeon preparatory to bleeding. So long as the ligature tightly compresses the arm, the blood will continue to flow from the opened vein. It is true that very rarely are the clothes drawn so tightly around the waist as the surgeon draws the ligature around the arm ; but an amount of pressure that the person would hardly be sensible of, or regard as injurious, will have a very evil effect if applied at the period of menstruation. By its specific gravity alone, the blood tends to remain in the lower part of the body, requiring all the powerful action of the heart to overcome this gravitating tendency ; so that under this condition any unnatural resistance, however small, is difficult to overcome. This blood that in tight-lacing gorges the lower part of the body, must either produce congestion in every susceptible organ, or find an outlet through a protracted hemorrhage at the menstrual period.

Every one is familiar with the common practice of tying up the arm by a sling to keep it in a horizontal position, when there is a wound or sore on the hand or forearm, to prevent the blood from gravitating excessively to the wounded part, and by its abundance causing inflammation in the wound.   The operation is purely a mechanical one, and is strictly analogous to the best means of relief in too profuse menstruation, *i. e.*, loosening the clothing, and reclining on a lounge or bed.   There would be just as little sense in applying styptics or other remedies to arrest this hemorrhage without abating first the principal cause, as there would be in applications to the opened vein to stay the blood while the ligature was still bound around the arm.

Another frequent cause of profuse menstruation among married women is the excessive sexual indulgence of their husbands.   This is indeed often an originating cause rather than a continuing one, as profuse menstruation tends to weaken the sexual desire.   An anxiety to gratify the husband, however, too often induces a submission to his embraces, when there is no desire on the part of the wife—a condition more injurious to her, so far as this complaint is concerned, than when the desire is mutual.

Thousands of wives, whom the world deems happy and comfortable in their marital relations, pass their lives in a state of misery and suffering from this cause. Their maternal organism, irritated and inflamed by the too frequent caresses of their husbands, affects not only their bodies but their minds, and fretful and feeble, hysterical and peevish, they pass cheerlessly through life, not well enough to live, nor yet sick enough to die. A wife suffering from profuse menstruation, or, in truth, from any complaint of the maternal organs, can not expect relief till this disturbing cause is restrained.

Another frequent cause of excessive, as also of painful menstruation among unmarried women, is solitary abuse of the maternal organs—a habit which is also the prolific parent of many other bodily complaints. A whole volume might be written on this subject—indeed, such are written, and it is not quite within the scope of this work to add to that literature. It needs only the remark to fathers and mothers, who may read this, that they should not remain ignorant of, or overlook the terrible blight, worse than death itself, that will surely overtake their daughters, and their sons too, who may contract this evil habit.

Having been led to say so much on this subject, the reader may naturally look for something
in the way of cure, or at least alleviation of the
complaint. This might appropriately be left to
the family physician, but having known instances
where mothers applied to these on behalf of their
youthful daughters just entering womanhood,
when he pooh-poohed the matter,—said "they
will be all right when they get older," or " when
they get married," or " when the function is once
regularly established," etc., a few words of advice may not be amiss. These will be hygienic
rather than medicinal. The main reason why
physicians' remedies so seldom cure this complaint is that the original exciting cause continues. The first thing, therefore, recommended,
is a careful study of the inducing causes for the
complaint, and an abatement of these. If it be
clothing worn too tightly around the waist, this
must be made loose—very loose. Violent exercise just before and at the period must be avoided, as dancing, running, horseback-riding, etc.
The bowels should be regulated, if necessary, between the monthly periods, so that all purgatives
may be avoided at the period. As much as possible the patient should lie down at the time

completely in a horizontal position, and not
merely recline in an easy-chair. Freedom from
all excitements of every kind is enjoined. These,
with the physicians' ordinary remedies, will materially help the sufferer.

Above all, she should not conclude, because the
alleviation seems slow, that there is no cure, and
no use in trying. The complaint, like many
other ills—corns for instance—is hard to eradicate entirely, but it may be brought into such
subjection as to be shorn of its principal evil
effects.

It is worthy of grave consideration whether
much of the sexual disorder of the present generation of young married women is not due, as
an inheritance, to the use of the poisonous pills
and drugs so openly and widely advertised thirty
or forty years ago for preventing the too rapid
increase of children in the family. That they
were freely purchased is evident from the noted
prosperity, financially, of the vile decoctors who
compounded and sold them; one of whom, dying
some years back, was reputed to have counted
her ill-gotten gains by the hundreds of thousands of dollars; each dollar of which probably
indicated a customer and a victim. No mortal

can conceive, only the power of Omniscience can estimate, the baleful consequences to the lives and health of the women of this country resulting from this one woman's nefarious business alone.

The "Society for the Suppression of Vice" is worthy of all praise for their successful efforts in stopping the open advertising and sale of these so-called remedies for female complaints; whose real purpose was indicated in the appended caution, that "they must not be taken during pregnancy as they would invariably produce miscarriage, though no other ill result would follow if by accident they were taken at such time." But there is a great work still to be done in this line by that Society or by other good people. It is quite notorious that large fortunes are made in the present day by as unscrupulous people who manufacture and sell these vile pills and potions, but who more shrewdly advertise them for curing "suppressed menstruation," "expelling tumors from the womb," and other as mythical female diseases, while the known real purpose in their use is to cause the abortion, at each recurring period of menstruation, of any conception that may have intermediately occurred.

No doubt these drugs are often used by wives
in blind faith in their alleged freedom from
everything prejudicial to general health, as always claimed by those who sell them. But wom-
en ought to know that any and all medicine
used to bring on the function of menstruation
when conception has once occurred, or, in plain
words, to produce miscarriage, is necessarily
deadly in its influence on the general health,
no matter what may be advertised about its innocent character.

The conception of a child is a natural obstruc-
tion to menstruation—in fact, a stop to it; and
this can not be removed or expelled (as some
seem, by the use of these remedies, to think) in
the same way that obstructions are removed
from the bowels by a purgative,—only requiring
stronger or special remedies. The womb is further removed from the direct action of medicines
taken into the stomach and passing through the
bowels, than any other organ in the body, and
all such remedies have to act indirectly. Necessarily before a drug can have this indirect action
with sufficient power to force the womb to open
and yield up its treasure, it has its direct and
more powerful action on the body, always poi-

soning the blood, and depressing the vital powers to the endangering of the general health,
frequently causing death itself before the con-
templated action on the womb is reached. The
conditions are like forcing a fruit-tree to untime-
ly cast its fruit by shutting off all moisture from
it, and laying its roots bare to the sun and wind.
Possibly before the tree would actually die the
fruit might drop off. But in vain would you
then supply moisture and cover up the roots;
the tree might languish out a few years of useless existence, but its value would be gone.

It is in no spirit of special animadversion that
this subject is dilated on. The remarks must be
general, as the extent of the evil can only be sur-
mised or inferred. But in many families, particularly among the educated and refined, only
two or three children are found, sometimes years
apart in their birth, where in the ordinary course
of marital life there should have been a dozen.
Of course it is possible that the limited number
are due to a Shaker-like continence in the married pair, though our common experience will
not readily believe this. It may also arise from
the Oneida Community practice of an intercourse
which avoids conception. But neither of these

5*

would bring customers to the compounders of
the abortive drugs in such great numbers as their
prosperity shows do come.   Hence by a process
of reasoning we arrive at the conclusion, that the
practice of avoiding maternity by the use of such
drugs is still wide-spread, and it is therefore a
fair surmise that this not only was, but still continues to be, a prolific cause directly, or by inheritance, of many of the deadlier forms of feminine disorders, which make their victims' lives
a burden to themselves and others ; fitting them
for, and reconciling them to, an early death.

In conclusion of this unpleasant subject it
need only be said by way of advice, that the
wife who thus keeps down the number of her
children not only brings on herself disease and
premature decay, but entails the sad consequences
of her folly and crime on the innocent children
she may later bring into existence.   For it may
truthfully be said, that a woman who once induces a miscarriage in this way, forever after in-
capacitates herself to bear healthy offspring.

CHAPTER IX.

THE editress of one of our popular periodicals
in some criticisms on the management of infants
very truthfully and forcibly remarks: "But a
great part of the children that are born nowadays are *not* good average healthy children.
They are children of deficient brain power, of
diseased nervous systems; children begotten of
tobacco-smoke, late hours, tight-lacing, and dys-
peptic stomachs. The father has put his son's
brains into his meerschaum, and smoked it out;
the mother has diddled and dribbled it away in
balls and operas. Two young people come together, both of them in a state of half nervous
derangement. She can not live without strong
coffee; her hand trembles and she has a sinking
at the stomach when she first rises in the morning till she has taken her cup of coffee, when

she is primed for the day. He can not study or read, or perform any real mental labor without tobacco. Both are burning life's candle at both ends ; both are wakeful and nervous, with weak muscles and vibrating nerves. Two such persons unite in giving existence to a poor helpless baby who is born in a state of such diseased nervous sensibility that all the forces of nature are a torture to it. . . . . What such children cry for is neither cold or hunger, but irrepressible nervous agony, sometimes from fear, sometimes because everything in life is too strong for them and jars on their poor weakened nerves, just as it does on those of an invalid in a low nervous fever."

The influences of tobacco, strong coffee, and late hours in enfeebling our race are not exaggerated in the foregoing. Indeed, the writer might with propriety have largely extended the list, for unfortunately evil indulgences are not limited to these few. And besides the committed evils there are many evils of omission which also lead to fragility and weakness. These last are indeed the most common among a large class who are carefully circumspect in avoiding the fashionable vices and follies of the day—the class known as the religious world, who are models of sobriety and refinement. In their desire "to keep the body in subjection," and to suppress the animal instincts as base and sinful, they are in danger of rooting them out altogether. They would be angels and "freed from earth and earth-ly failings" before their time and while yet in the body, and so they but half do the work their Creator has given them to do in the world.

What is needed with this large and deservedly influential class of women is that the desires connected with maternity shall receive a higher respect,—that there shall be accorded to the fulfilment of the maternal duty a holier and less sensual idea. So that its legitimate and reasonable gratification may be honored and ennobled; so, too, that its suppression or wicked perversion shall be regarded as a sacrilege as deserving the curse of God as was the crime of Onan.

As in eating, there is a happy medium between fasting and gluttony which we readily recognize as the true rule; so there is a proper middle ground in the marital intercourse, between squeamishness on one side, and sensuality on the other, which should be our path, both in our own course and in the training of our children. One great reason why wives are so often ashamed nowadays to have children is, because of the physical and mental disqualificatións of the few they do bear. There is nothing in their infancy or maturity to be proud of, unless, as is too often the case, alas! we are proud that they have inherited a concentration of our own infirmities. There was never yet a right-minded wife who would not be proud to be surrounded with a round dozen of stalwart, clear-headed sons, and rosy-cheeked daughters; but when they come forth pale, stunted, scrofulous, and spine distorted, fit only for the physician's hands, one or two quite suffices to satisfy the mother's pride.

It is said that a family living in Paris will die out and cease to exist in about five generations, and from many observations of the writer, it may probably be truly said of families living in all large cities,—the tendency of such places being gradually to debilitate and impair the race. But this will be found true only where the sons continue to intermarry with the daughters of contemporary city families. The introduction from time to time of healthy, well-developed country girls as wives would retain the name and lineage in existence.

Almost every person of mature years who is extensively acquainted among city families, can recall many instances among his acquaintance where the elder members of the family born and reared, perhaps somewhat roughly, in the fresh, healthful atmosphere of the country, and transferred to the city at a mature age, continue hale and hearty at seventy or more years, with a numerous progeny, who reared in the city fall much below their parents in general health and strength, and in the third generation are so feeble and delicate that half of them die off before their grandparents. If such families be looked into closely, it will be found that there is a fair proportion of sons among the children of the first generation, fewer among the children of the second generation, and a large surplus of girls in the third generation; so that in the fourth, or at most fifth generation, the males have all died out of the family, and the family name becomes extinct. The exceptions to this in the birth of an occasional son in the later generations may generally be traced to some specially advantageous condition of the mother, either herself fresh from country life, or if descended from a contemporary city family, that she has enjoyed some special opportunity for the restoration of her family health and strength. There is a consoling thought in the fact that if life in the city, with its impure air and excitements, can reduce the health and strength of the race, a return of even the weakened ones to a more favorable condition of existence can again in a generation or two restore the race to its pristine vigor, so that though the transgressions of the fathers and mothers may be visited on their children, the children may modify the infliction, and by a judicious regimen and training eradicate from their lineage the hereditary maladies, and need not, perforce, transmit to their offspring the physical ills they themselves inherited—at least not to their full extent.

We will now turn to the bearing these reflections have on the main subject of the book.

A wife whose children have been only or mainly girls, and who is desirous of having boys, should carefully study the physical character of her ancestors as well as her own. If those on the mother's side especially, have been feeble and prone to give birth to girls, she may regard herself as clearly under the same hereditary influences, which only time and attention can hope to eradicate. If, on the other hand, her ancestry have not been specially given to the production of daughters, she should look to herself and see whether by a too sedentary and idle life, she has not fallen away from the pristine strength and vigor of her ancestors. If this is the case, she may hope to recover the vigor necessary to the conception of boys, by a change of scene and habits; getting away into the country, and taking active outdoor exercise. By restraining at the same time all sexual and other indulgence which may have tended to her enfeeblement, she might hope in a twelvemonth or so to regain her strength and be capable of conceiving male children. Of course, while this disability exists, whether hereditary or personal, child-bearing must be postponed; by an entire avoidance of intercourse with her husband if necessary, and certainly not by the use of poisonous and life-destroying medicines, than which nothing can be more destructive to the strength of the female organs of generation.

It may seem to some readers that there is an indelicacy in this sort of preparation for an act that is ordinarily unpremeditated, and not resulting from forethought or intention. But if the right view be taken a different conclusion will be reached. We have as a people grown exceedingly wise in the propagation of fruit, vegetables, and farm stock, and can discourse scientifically upon the various processes involved, and of the great benefit to the human race made by science and observation in these minor things. But when we come to speak of the improvement of our own race in the generation and conception of the embryo man, we are met with the idea that the subject is indelicate, and should not find advocacy among a cultivated and refined people. Now there is a phase of the subject which is indelicate,—that which brings it forward as the topic of social conversation and as promotive of erotic thoughts simply for their own sake; but when it is confined to the sacred privacy of the marital chamber, the mind should rise above the thought of personal enjoyment, to the contemplation of the agencies that bring into being an immortal soul, and not only should we desire it, but should use all our reason and all our knowledge to have this immortal soul brought into the world under the most favorable circumstances.

It is a somewhat prevalent belief that male

children are more liable to be aborted than fe-male, and the statistics show that this belief is founded in truth. A male child requires in the mother an amount of vigor and strength in its conception, and subsequent fœtal growth, much beyond what a female requires. Hence there is a tendency to overtask the womb, which if not suf-ficiently strong, acts in the way nature ordinarily adopts in similar circumstances in an overloaded stomach, and throws off the burden. Nature only bears within limits. Up to a certain point the plant or animal may be taxed for the pro-duction of fruit or offspring. When we go be-yond this point without increasing the strength of the parent plant or animal, and fruit sets or conceptions take place, which too much exhaust the parent, such fruit drops off prematurely until the quantity is reduced to the limit that can be perfected, and with the animal abortion ensues—Nature in both cases intervening by the primal law of self-preservation to relieve the parent. Even when the fruit or offspring by much care and precaution sometimes reaches maturity, it seems to lack some essential element of life, and is evanescent—the fruit soon decays, the offspring rarely survives to propagate its race.

When, therefore, a wife miscarries with male children while she carries females the full period of gestation, the reason for it will be found in her sexual ardor being strong enough to conceive males, while physically she is not strong enough to perfect the conception into a living healthy child, and needs to adopt measures to increase her health and strength at such times, and especially must avoid everything tending to excite the reproductive organs; particularly must the cohabitation with her husband be avoided.

In the previous chapter reference was made to the evil of excessive sexual indulgence, as causing by its irritating effects a too profuse menstruation, but its evil effects do not end with this; these are wide-spread and ruinous on the wife's health as well as often injurious to the husband himself. It is incumbent on every husband to see to it, at least, that he is not indulging his desires at the expense of his wife's health. The control of this matter lies with him, as the wife is accustomed to submit to his caresses even when they are not gratifying to her; the contrary of which can not well occur. Doubtless, the most of husbands who are guilty of this injurious course toward their wives, are guilty from ignorance. There is, in truth, too much ignorance among all the community of the evil effects of unrestrained indulgence of the sexual passion. It is too much taken for granted that so long as nature induces a desire of this kind, one can not be hurt by its indulgence. This is a great mistake. The seeming natural desire may become as vitiated as the thirst of the drunkard for spirituous liquors, and its gratification as dangerous and debasing.

The writer frequently meets among his acquaintance married people who are visibly sufferers from this cause. They are always ailing; the husband can not half attend to his business, he has a headache, or a fever, or a cold, or sickness of the stomach, or bowel complaint, and his wife more or less in the same condition. The close observer sees in their lustreless eyes, their sodden and greasy faces, and their trembling hands, evidences that an almost nightly indulgence is kept up of the pleasures of the marriage-bed, which is the origin and cause of all their ailings,—sapping, as it does, the very foundations of their vitality.

In this western continent this desire seems morbidly active, whether from the climate, or habit of living, or both, is uncertain. It is no doubt largely the cause for the thinner and sparer forms in this country, as compared with the plumper and rosier-faced inhabitants of the Eastern hemisphere. Its effect is seen on our married immigrants, who, reaching our shores plump and rosy-cheeked, are found five or six years later with much of this plumpness lost. Acquaintances of this character have admitted to the writer, an increased desire for these marital pleasures, mainly attributed by them to "higher living" here than they were accustomed to in their earlier homes. It is said that there is a larger share of oxygen in our atmosphere than in the older and more populous countries of Europe. If so, this may be an exciting cause for the increased sexual desire. Whatever the cause, whether dietetic or climatic, it seems to need that especial restraint should be exercised, so that our bodies are not wasted, and our race enfeebled by excessive indulgence of this passion.[1]

Our physicians need to take this subject more into consideration than they seem to, in the investigations of the causes of disease. It lies at the bottom of very many of the ailings that humanity groans under, especially those called

1 See Appendix, Note F., page 159.

chronic ; which are only so because the patient
is chronically a transgressor in this particular,
and by his unwise indulgence is continually
undermining and breaking down, as fast as
nature, aided by medicines, restores.

Before closing this practical part of the subject, relating to the application of the " physical
law," I bethink me that some man seeking a wife,
and desirous of a fair proportion of sons after
his marriage, may ask, Is it not better to select
for a wife a woman who has not these disabilities,
rather than to try to cure them after marriage, in
one who labors under some of them, and what
are the characteristics by which such a selection
can be made ?

The answer is, There is no better way for the
restoration of a family line, and of the race at
large, than by the intelligent and judicious selection of the fittest women for wives, and the
future mothers.   As there is a surplus of unmarried women in most of our towns, this is not
difficult.   It may be sad for those of the sex who,
being feeble in frame and delicate in health, will
be thus left to single-blessedness by this selection.   But it is more sad to think of these, as
wives and mothers, continuing to entail all their

weaknesses on their offspring generation after generation. Indeed, the thinking young women of to-day, who know themselves to be in this debilitated condition, do not hesitate to prefer a single life, unless, by chance fortune, they find a congenial husband, who prefers that his wife shall not be the mother of any children.

What the special qualities are to be selected in a woman for a wife, may be catalogued as follows:

She should, in the first place, be of good, sound health, and free from all bodily ailings, especially free from all complaints and irregularities of the maternal organs.

She should be not more than five years younger than her husband,—a desirable age is between 20 and 25 years.

She should be in height five feet to five feet six inches, and not more than four inches less in height than her husband.

She should measure, over a single light undergarment, *at least* 36 inches bust measure under the arms, 26 inches waist measure, and 38 inches around the hips, and weigh not less than 140 pounds.

She should have a good, strong growth of hair,

and bright lustrous eyes, both of any color the
husband likes, and a fresh, clear countenance.
Dullness of the eyes, and a pale, sallow, or pimpled
face frequently indicate something wrong in the
sexual organs.

She should have strength sufficient to readily
put up a 25-pound dumb-bell in each hand,
alternately, to arm's length over her head a dozen
times at least, and to take a five-mile walk, out
and back, every day, without special fatigue.

With a wife answering to this description, a
husband will have to be endowed with more than
ordinary virility to beget any daughters. And
such a wife will leave to his children as great a
fortune as he can possibly leave them, though
his wealth be counted by the million, for she
will endow them with good health and strong
constitutions; provided, of course, that there
are no serious defects in his own constitution for
them to inherit.

In conclusion of this subject, it is hoped that
this " physical law" governing the conception of
sex, and its method of operation, have been so
clearly and fully described, that the reader under-
stands them thoroughly. But general principles
could be given at best. The subjects, in their

6

details, are profound studies, requiring volumes
to fully compass them.   The causes which tend
to weaken the maternal organs of the woman before and after marriage, considered alone, would
be a life's study, embracing as they do, not only
the special maladies of the sex, but to a great
extent the reflex action of most of the maladies
to which the human frame is liable.   The object
in what has been written about these is to put
the reader on the track for investigation of each
individual case, and have it managed on the
general principles laid down in medical books, or
by the treatment of the family physician.   It is
unfortunate that too many of these estimable
men pay so little attention to the origin and cause
of disease, being generally content to treat the
more prominent symptoms, and more distressing
effects, without going back to get rid of the active
cause of the malady.

Possibly, where the originating cause of the
ailing is an evil habit, it may not be strictly in
the line of professional duty to revert to it.   The
uncovering and exposure of evil habits, which
the patient preferred to have hidden from all the
world, has so often been the cause for discharg-
ing the physician from further attendance, that

some allowance must be made for the indisposition to search into the mysteries of the case, further than the patient chooses to reveal them.

It is believed that, with the guidance herein given, no woman need be at fault in coming to the conclusion as to the cause of any sexual disability she labors under, nor ignorant of the best methods to obtain relief.

And should my investigations, and the publication of this book, be found to practically influence the production of the sexes in the human family, so that a much larger proportion of boys shall be born, as I believe there will be, if the teachings be faithfully followed, future generations will not see, as in the present superabundance of unmarried women, so near a possible fulfilment of that prophetic period, mentioned in Holy Writ, when " seven women shall take hold on one man, and say, We will wear our own apparel, and eat our own bread, only let us be called by thy name" (Isaiah iv. 1).

The greater tendency of boys to die off at an early age than girls, has been frequently mentioned in connection with the statistics and elsewhere in this book. The subject demands more than such indirect mention, and might have a chapter to itself, only that it is not in the line of this treatise, as shown by the title, though it is fairly in the line of the author's purpose, which may be broadly stated as an effort to secure, in coming years, a greater proportion of men in the population. What may be succinctly said on this subject seems to find its most appropriate place here.

It was to the author not only a sad but a remarkable fact, that the three sons born to him, as mentioned in the introductory chapter, all died in early infancy, while the girls lived. Even the one born intermediate between the second and third son, peculiarly delicate and feeble, and with less promise of living to maturity than either of the sons, came safely through all the usual infantile complaints and grew up to be a fairly strong woman, while each of the sons in turn succumbed to what seemed at the time much less potent maladies ; neither one living till the next son in order was born. It was reflection on this fact, and the observance of similar instances of the death of sons among his acquaintance, that led to the investigation of this collateral subject.

It is, therefore, not enough that by following the instructions in this treatise sons are born in families where daughters in excess have been the hereditary or individual rule, but these sons must be specially cared for after birth. As a rule the mothers who have mainly daughters are not good wet-nurses for their children—at least not for their sons. There is a natural relation between the two organisms in the maternal body for child-bearing and child-nourishing that when the first is too weak to conceive male children the second is likely to be deficient also. And it is specially desirable that the sons born of such wives should have a better nourishment than the mother is able to supply.

It is not always with such mothers that the supply of milk is limited in quantity, but it is often inferior in quality when it is abundant. Among the varied food a mother of this kind eats, there will be every few days something that she does not digest well—or as she may say, "that disagrees with her." This influences for the time the quality of the milk secreted, producing intestinal difficulty in a feeble child, and when repeated again and again eventually produces an irritated condition of the child's digestive organs, through which its vitality is in some degree exhausted, and it falls a ready prey to disease.   When had the mother been endowed with a better digestive apparatus her child would have continued strong enough to resist the attack.

This condition of the mother's digestion should therefore have full consideration on the birth of a son among a family of daughters.   If there is any defective nutrition the babe should have a healthy, robust wet-nurse at once.   And even she should be cared for in regard to diet, as most women of this class are grossly ignorant as to what they should eat and what not for the best interests of the child.   In her selection - something more than merely a full supply of food for the infant should be looked after.   It is quite important with her also, to see that its quality is good, and that she by her good digestion can maintain its good quality.   To this end unless she is known to be a healthy woman this condition should be examined into by a competent physician.   Any effort to supplement the mother's limited flow of milk or its inferior quality by artificial food during at least the first year of the infant's life is likely to be fatal to its existence.

The advantage of pure country air for all chil

dren is widely recognized, but it is invaluable in
rearing healthily these tender sons of feeble
mothers. They should always have the benefit
of this, at least in the warm months of the summer, when outdoor exercise is specially de-
manded.

CHAPTER X.

THIS law is as applicable to the animals of the farm as to the human race. Indeed there are less difficulties in the way. The equality of condition and living with both sexes is the general rule, unlike that which exists among men and women, and there are not so many disturbing elements to ward off or subdue.

If male offspring among these are desired, the female must be vigorous, in the best condition of health and strength, in the full prime of her life, and not exhausted at or just before the union with the male by overwork, or worried and fatigued by being driven. She should meet the male at the period of greatest heat, and only one connection should be allowed. More than that naturally weakens the female's ardor, and if the conception occurs from the later connection, there will be a greater probability of its being a female.

(128)

When the females are young, they are often timid and frightened at the presence of a strange male, which tends somewhat to detract from the force of their passion ; and it is well in such circumstances to permit a social intercourse between them apart from other animals for a few days prior to the period of heat, so that she may become accustomed to his presence, and be willing to accept him when the proper time comes without disturbing fears. The male ought to be in a condition of strength and lustiness inferior to the female, at least he should not be in a superior condition. By condition is not meant fatness, but bodily health and vigor.

Experience has taught the cattle-raiser that among farm-stock where the male and females range at large together, mature females coupled with very young males, are more likely to produce male progeny, while young and immature females coupled with mature males, are more likely to have female progeny. This is in accord with the physical law enunciated. The young and scarcely mature male does not in his ardor rise to as high a degree as the mature female. Similarly the young female has not the intensity of ardor of the mature male ; the re-

6*

sult being that the mature female has male offspring, and the immature female has female offspring. Of course when female offspring are
desired, the opposite course must be pursued.
The female must be kept lower in condition, and
the male so nourished as to be in a higher condition of lustiness.   If as is usually the case one
male serves a number of females, they should
not be brought to him in too immediate succession, while he is exhausted by his previous service.   A day or more should be allowed to inter-
vene.   Even a longer period is desirable if the
female is in especially good condition, as many of
the finer blooded cows are kept. With these
when a heifer calf is desired, as it usually is, it is
well to secure in advance the services of a vigorous bull whose ardor will be strengthened by at
least a week's absence from a cow. This might
be done by letting the cow pass her first period
of heat, making such an arrangement for the
bull's services at the next period; keeping her
during the intervening time on a spare diet.
Driving her some distance, so as to exhaust her
strength and thereby possibly some of her ardor,
just before introducing her to the bull, would
have a beneficial influence in inducing the con
ception of a female calf.

It must also be understood that low feeding and high feeding, to be of the use here described, ought to be of some continuance, and not merely a temporary expedient of two or three days' duration. It is more likely to be efficient for the purpose when it is the natural and ordinary condition of the animal year in and year out. The sexual desire of the female during lactation can hardly be sufficiently influenced by the best and most nourishing food if continued for only a short time, as its effect is mainly to increase the flow of milk : though when a male is in good condition, and only likely to be somewhat exhausted by too frequent service of the females, a stimulating food in the intervals of rest even for a day or two will greatly aid in restoring his ardor.

The manuscript of this book was already in the hands of the publishers when they called my attention to an editorial article on "Controlling Sex in Animals," in the *Rural New-Yorker* of July 5, 1884, in answer to the inquiries of a correspondent. Among several theories or practices said to be found more or less effective, the editor mentions "that a Texas gentleman, Mr. D. D. Fiquet, claims that he has repeatedly controlled the sex of domestic animals by food alone. For instance: when he wished a heifer calf, he fed the cow light cooling food for some days before putting her to the bull, and at the same time fed the latter with rich heat-producing food; but when he wished to obtain a bull calf, he reversed the practice."

His practice carries out in an empirical way the theory of this treatise—that to produce female offspring the male must be in a more vigorous sexual condition than the female; and reversely, to produce male offspring the female must be in a more vigorous sexual condition than the male. But there are other ways than by this special feeding that these conditions can be brought about, as the reader will have seen.[1]

1 See Appendix, Note G., page 160.

CHAPTER XI.

IT is scarcely to be expected that in this scientific age any mere empirical conclusion will be accepted as a natural "physical law" by intelligent thinkers unless somewhat supported by a reasonable scientific theory explaining the *modus operandi* of the action that leads to the conclusion.

The physician may find from his many experiments that a certain drug is a specific in a given disease, but he always somewhat doubts its universal applicability to all ages and sexes and conditions of humanity, till he can fathom the method of action of the drug, and be able to tell how it produces the curative result. Now the writer is not a professed scientist, but having reflected considerably on the subject of the influence that determines the sex of the embryo, as also on the wider field of the tendency or inducement or desire that brings together the two sexes, both of plants and animals, in the reproduction of their varied kinds, presents the following as a theory to account for both phenomena.

There seems good reason for concluding that this mysterious influence which brings together the two sexes in reproduction is one of the phenomena of electricity ; known in some of its phases as "animal electricity," and that it is governed by the somewhat mysterious laws of that fluid.

If we accept the current theory that electricity is composed of two fluids or influences, which by heat, friction, or chemical action can be separated and which have such an affinity for each other as to readily unite again, we can find several analogies to this action in the process of generation in both plants and animals.

It would be out of place here to discuss at length the character of the electric fluid. It is itself a mystery quite as great as that of the cause of sex in offspring ; and it must suffice to say, simply, that while one theory of it is that the fluid is one simple element whose twofold manifestations indicate merely its presence in excess, or its absence in a measure, from which the well-known terms of positive and negative electricity are derived; the generally accepted theory regards it as a compound of two elements to which the terms vitreous and resinous have been applied, because a vitreous substance, as a piece of glass on being rubbed with a silk or woolen cloth takes on uniformly one element— the positive or vitreous, while a resinous substance, as a stick of sealing-wax, on being similarly rubbed takes on as uniformly the other element—the negative or resinous.

Applying these elements to account for this universal desire between the two sexes for union in the act of reproduction, if when animals and plants arrive at their several periods of puberty or maturity, and are replete with vitality, the male organs secrete or eliminate positive, and the females negative electricity, we have a universal affinity prevailing in all species influencing each sex at such times to fly to meet its complementary or opposite sex with all the force of magnetic attraction. Among the lower organized beings, as plants, fishes, etc., the mere contact of the products of the male and female organs, as the pollen and ovule of plants, the milt and ova of fishes, is sufficient to produce a mutual electric discharge, at which time impreg

nation occurs. In the higher orders of animals
this discharge does not take place until after a
certain amount of excitation or friction together
between the diverse organs of the sexes, wherein
they exert somewhat the same influence on each
other as the glass and pad of an electrical machine, the testicles on the one side, and the ovaries on the other, acting as temporary reservoirs
of the accumulating electric fluids until so replete that they can no longer retain the accumu-
lation, when a mutual electric discharge occurs,
carrying with it on the part of the male the
secretion or germ natural to his organs. This
germ must not be confounded with the magnetic
fluid itself, as the electric discharge can take
place without any germ, and the germ exist
without any electric discharge.

It will be seen that this theory regards the
difference in sex as founded on the different
electric influence of each, the male being always
positive, the female negative ; or using the other
terms, the male developing vitreous, the female
resinous electricity; different species having something peculiar in the character of the electricity
they generate, so that each species throughout
all creation, has more inclination for the oppo-

site sex of its own species than for a different
one: akin in this respect to the affinities found
in many chemical substances.

It is not meant that this sexual electric condition is found to such a great extent, in the
muscular parts of the body generally when in
repose, but that it exists mainly in the generative organs at times of sexual excitement.
Though circumstances induce the belief that it
does often extend to all parts of the body, so
that a man in the full vigor of his life can tell by
certain sensations that a hand he clasps in the
dark is that of a woman and not of a boy, particularly if the woman is at the time possessed
of some amorous desire.

It is not very long since, in a divorce case in
New York City, that the wife while admitting
her guilt, her paramour being a young man living in the house, a near relative of her husband,
pleaded in defence her husband's condonation
following on her confession to him and other
near relatives. In this confession, she sought to
extenuate her crime by asserting that it was not
premeditated, but that being in the young man's
chamber early one morning on some household
errand, he grasped her by the hand, and straight-

way such a thrill went through her whole body that she had no control over herself. Those who read the account, naturally smiled at the simplicity of the excuse; but it is not hard, if this theory be correct, to believe that what she confessed was actually the truth.

The warm clasp of the hand between two lovers which thrills the hearts of both is made much of by the poet and novelist; but sagacious minds will strive to learn the physical cause for it, even though the knowledge may spoil the romance.

In what is called chemical electricity there is another characteristic strongly supporting this theory. In most of the works on electricity a list is given of some fifty to sixty elementary substances, each of which is positive to its next neighbor above on the list, and negative to its next neighbor below. At the extreme positive end of the list is potassium, the most powerful alkali; at the other, or negative, extreme is oxygen, the most powerful acid. And so far as the substances have known alkaline or acid qualities they seem regularly graded along the list according to these chemical qualities. Thus sodium, a slightly less powerful alkaline than potassium,

is negative to it, but positive to lithium, the next
on the list, seemingly because it is more alkaline
than lithium. Chemical nomenclature of compound neutral substances is largely founded in
this, and the proportions of the electro-negative
element and the electro-positive element in the
mixture fixes its distinctive name.

Further investigation in this path will possibly show that chemical affinity is but another
name for electrical affinity, or mayhap that all
electric action is only the chemical action of al-
kalines and acids. It is foreign to the subject of
this work to inquire into that ; the suggestion is
only thrown out, that in following this clue it
may be eventually revealed to us what electricity
really is.

It is a peculiar circumstance in support of this
theory that long ago the able chemist, Dr. Franz
Simon, in investigations of the chemical qualities
of the various secretions of the human body,
found the menstrual discharge of woman to be
quite acid in character, while the semen of man
was as decidedly alkaline ; being the two chemical conditions which, according to the list mentioned above, would naturally generate in the
woman's sexual organs a negative electric con-

dition, and in the man's a positive electric con
dition. And it is a fair inference that the secretion of the male and female organisms, both of
animal and plant life, are alike alkaline or acid,
and generate the positive or negative electric conditions according to the sex. There is another,
perhaps trifling, analogy between the action of
the electric fluids and of the sexes in reproduction that corroborates this theory. Careful experiment has proven that when there is a mutual
electric discharge between a positively electrified
body and one negatively electrified by means of
a conductor, that the positive electricity travels
along the conductor a much greater distance
than does the negative, so that the point of
junction of the two elements is comparatively
near the negatively electrified body. Now this
peculiarity seems to be universal in the masculine character. The male goes further to meet
the female, than does the female to meet the
male. Shall we say this is simply because the
desire of the male is greatest, and rest there?

The diverse desires of the two sexes seem in
fact to be quite influential in the earlier evolution of organic existence. So far as the microscope reveals it, the first evidence of life is *re-*

production. The minute cell or bubble which to all our scrutiny is but a film of dead matter enclosing a gas or invisible fluid is seen to reproduce itself by the growth of a similar cell from out itself, indicating a bi-sexual existence and action prior to any evidence of the process of nutrition. From this the first and lowest condition of plant life where the sex organs are entirely hidden, they seemingly develop by innate power to the next or higher group—the so-called hermaphrodite, where the sexual organs are separately visible, as in the stamens and pistils of most flowers. Proceeding still further along in the innate development, we reach the monœcious group, as the maize or Indian corn, where these organs through their varied aspirations or tendencies are become entirely separated, though both still on the same plant. Still further on we reach the diœcious group, as the ailanthus, where this mysterious influence of sex has produced a total separation, the whole plant being either male or female, as exist in the higher animal life. From observation of these tendencies it would almost seem as if reproduction was the primary purpose of creation, and that the usefulness of the individual outside of this is second-

ary : even as the fruit and vegetable food generally, that we cultivate and eat, is but a provision
stored up by the parent plant to protect and aid
the germination and growth of its offspring.

To understand clearly how electric influences
can determine sex, attention must be given to
two well-known phenomena of electric action.
(1.) That every electrified body induces an opposite condition of electricity in every susceptible non-electrified body with which it comes in
immediate or close contact. This is known as
induction, a common illustration of which is the
taking up of needles or small pieces of steel or
iron by a common bar magnet. The positive
end of the bar being able to attract such an article only when it is in a negative state, produces
when first applied to the end of a needle, by induction, a temporary negative condition in that
end, after which the attraction between the two
causes the needle to attach itself to the bar. The
equilibrium of electricity being disturbed in the
needle, the end remote from the bar becomes
positively electric, and in its turn if presented to
another needle induces a negative condition in
the end it approaches, and the second needle
then attaches itself to the first, so continuing

adding needle to needle, till the power of the magnetic bar is exhausted. This induction is also seen in charging a Leyden jar where the inside being charged with positive electricity, an opposite or negative condition is induced and appears on the outside of the jar, the induction acting to produce this through the glass of the jar, though the glass being a non-conductor, the two opposite elements are retained separate until mutually discharged by a conducting wire.

(2.) The phenomena known as the residuary charge. Thus, if after the Leyden jar be partially charged, it be insulated so that the negative electricity can not be collected on the outside from adjacent objects, and the interior charge be continued, there will be a larger quantity of the positive electricity in the interior than there is of negative on the exterior. If now conduction be established between the outside and inside of the jar, a discharge will take place of all the negative electricity and an equivalent portion of the positive, leaving, however, a residuary charge of the positive, which if the jar be now uninsulated, will induce another flow of negative electricity to the outside. So that if by any means two bodies dissimilarly charged with electricity be brought into contact, that having the greatest charge will neutralize the other, and have a residuary charge left to influence by induction the opposite condition in another body brought into its near vicinity.

It is important that these well-known phenomena of electricity be clearly understood now, to understand how they can act to influence the sex of offspring, especially the first one—the natural tendency being, with those not familiar with the subject, to suppose a positive condition of electricity will induce a positive in surrounding objects, and a negative condition induce a negative.

These being borne in mind, we can see there must be always in the male parent an influence tending to induce the negative or female form in the germ he produces ; it being to him when once formed and passing out an independent existence, or foreign object. While on the other side the opposite electric condition of the female parent would similarly induce a positive or masculine form in that part or portion she furnishes toward the making up of the embryo. But the female parent may be so feeble in her sexual desire, or negative electric condition, that she has been unable to exert much electric influence on the part she furnishes, hence at the instant of the voluptuous paroxysm, when the two electric influences or forces unite, and conception ensues, she has been unable to give her share of the embryo, by the inductive influences of her sexual organs, a sufficient positive condition to over come the negative condition of the germ presented by the male parent, so that as between the two electric influences commingling in the embryo, that furnished by the male, and by him negatively induced, is superiorly charged, and the residuary charge remains negative, the embryo continuing and developing as a negative body or female. Whereas had the female parent furnished a larger supply of her natural negative electric condition at the time, so as to have induced a larger supply of the positive electric condition to the embryo, than was furnished of the negative by the male, this positive condition would have overborne the negative, with the result that the residuary charge would have been positive and the embryo taken on at once the male form. That there is a continuous electric action of this character in the sexual organs, two circumstances somewhat indicate.

7

One is, that wives who have had several children, boys and girls, can generally tell when pregnant with a boy, because during that time their sexual desires are excited to a much greater degree than when pregnant with a girl. The negative electric condition of their maternal organs is probably continually excited inductively by the nascent positive electric condition of the male fœtus. The second is, it is noticeable that the males of animals whose young are nourished and developed by a fœtal growth in the womb, have their sexual organs more or less drawn out of the body, while the females of these have theirs as deeply reverted. As if in the progress of development of these orders the organs of the males had been in the process of ages gradually drawn out by the mutual attraction of the different electric conditions of the mother and the male fœtus ; and in the female fœtus gradually reverted by the mutual repulsion incident to the same electric conditions.

The males of animals grown from the egg, and without the fœtal growth under the influence of the mother's sexual organs, are not drawn out, but remain in the body, nor are the female organs in these deeply reverted.

The views expressed in this chapter were at the outset denominated a *theory;* they might, perhaps, better have been called a *hypothesis* to account for the existence of the physical law governing the sex of offspring as presented in this treatise. Necessarily they are purely speculative, and if they can not stand the scrutiny of closer investigation, the physical law itself is not compromised thereby, any more than is the physician's specific affected in its curative properties when his theory of its action may be found erroneous.

1 See Appendix, Note H., page 173, and "ANSWERS TO OBJECTIONS," page 182.

A'PPENDIX.

NOTE A. (*See page* 26.)

In the *Popular Science Monthly*, Vol. XXVI., p. 323, there is an article by Professor W. K. Brooks, of Johns Hopkins University, reviewing some communications in the *Jenaische Zeitschrift*, by Carl Düsing, on the "Laws of Sex." The Professor remarks that Düsing has gathered by far the most voluminous statistics of any writer on the subject, some of which he gives. These show that proportionally, more girls are born in the city than in the country. Apparently from this is deduced the conclusion that " a favorable environment causes an excess of female births, and an unfavorable one an excess of male births." I should regard the city as an unfavorable environment for healthful life, and think our country life, where there is abundance of food and good, pure air, with such measure of exercise for the people as insures good appetites and good digestion, rather the most favorable, and there, statistics show that the proportion of boys born is much greater than in cities.

It is evident by the statistics that it is city life, with its ease and relaxation, the usual accompaniment of wealth, found oftener there, which he calls "favorable," and which tends to increase the proportion of female births. Mr. Düsing's supposition is, this result comes from some natural law to prevent close interbreeding, etc. Professor Brooks thinks this supposition not tenable. He concludes that in the unfavorable environment of the country, where proportionally more boys are born, "this result comes from a deprivation of food, and the increased birth of boys is an effort of nature to save the race from extinction." It is not obvious how the increased births of boys is to help that, admitting the want of food the cause for this increase.

That this result in city life does not follow from the above supposed cause, but that it is a consequence of loss of vital, and consequently of sexual energy from the life of ease and relaxation of the wealthier wives, as I have set forth in this book, is, I think, quite conclusively shown by the investigations of Francis Galton, F.R.S., as recorded in his work "Hereditary Genius," London, 1869. He says, chap. viii. (*On Heiresses*) :

"Of the peers who have married heiresses from the time of Charles II. down to the end of the reign of George IV., the unions produced per hundred,

208 sons, 206 daughters. While the marriages with women not heiresses produced per hundred, 336 sons, 284 daughters. Of the heiresses 22 per cent. had no son, 16 per cent. only one son, and 22 per cent. only two sons. While of the wives not heiresses only 2 per cent. were without a son, only 10 per cent. limited to one son, and only 14 per cent. limited to two sons. In many instances by the second and third generation following these heiress wives, the sons had all died out, and the peerage became extinct. Out of twelve peerages extinct eight are clearly traceable to intermarriage with heiresses."

Mr. Galton gives no reason for this condition except " the supposition that as the heiresses were often the only child in the family this infertility might be hereditary." Though he goes on to say: "An heiress is not always the sole child of a marriage contracted early in life and enduring many years. She may be the surviving child of a large family, or the child of a later marriage, or the parents may have **left** her early an orphan. We ought to consider the family of the husband, whether he be a sole child, or one of a large family. These matters should afford a very interesting field of inquiry to those who care to labor in it, but it falls outside of my line of work."

From these statistics of Düsing and Galton, I can not but be more strongly confirmed in my own conclusion, as set forth in this book, that this tendency to the increased proportion of girls arises from the softness and lack of physical vigor, so often found in women surrounded by wealth, and who pass their lives in luxury and indolence; a condition oftenest found in city life.

(*See also in this connection Note B.*)

NOTE B.   (*See page* 43.)

Doubts have been expressed as to the increased proportion of girls born in cities being so great or so general as represented. Though the last census enumerates over 290,000 more females than males in the states east of the Alleghanies, these mostly congregated in cities, and the last English census (1883) shows for London alone 467,887 more females than males, this condition is often wholly ascribed to the emigration of the males. To show that this cause is inadequate to account for this redundance of females, a table has been made up from our last census embracing all the cities with over 100,000 people, showing the total population by sex, and the school children five to seventeen years inclusive by sex also. - While there is no doubt that the males in the eastern cities are decreased and in the western increased in number by emigration, it is not probable that the proportion of the sexes in children of the school age are greatly affected by influences of this kind, as children of these ages rarely migrate except in company with their parents, and then both sexes go together.

The proportions of females to each 100 males are added in adjacent columns for convenient comparison. As the school children are tabulated in the census only by counties, the figures given are for the county in which the city named lies, wherever the county is named in the table, and this both for all ages, and for the school age.

City.	County.	All Ages.			School Age.		
		Males.	Females.	Females to 100 Males.	Males.	Females.	Females to 100 Males.
New York.....	....	590,514	615,785	104.3	150,051	154,731	103.1
Philadelphia..	....	405,975	411,195	108.7	104,793	106,764	101.9
Chicago.......	Cook..........	311,254	295,270	95.2	81,577	83,586	102.5
Brooklyn......	Kings	289,153	310,342	107.3	79,724	81,432	102.1
Baltimore....	....	157,393	174,920	111.1	42,676	45,398	106.4
Boston........	Suffolk.......	183,995	203,932	110.8	43,560	44,673	102.5
St. Louis.....	....	179,520	170,998	95.3	47,988	50,119	104.4
Pittsburgh...	Alleghany.....	179,114	176,755	98.7	52,934	53,196	100.4
Cincinnati....	Hamilton......	154,481	158,893	102.9	43,371	43,935	101.3
San Francisco	San Francisco	132,608	101,351	76.4	28,392	28,365	99.9
Buffalo.......	Erie..........	110,029	109,855	99.8	31,384	31,209	99.4
New Orleans..	Orleans.	100,892	115,198	114.2	29,448	32,081	108.9
Providence...	Providence....	94,921	102,953	108.5	24,194	24,645	101.8
Cleveland....	Cuyahoga......	99,005	97,938	98.9	27,867	27,913	100.2
Newark........	Essex.........	91,605	98,324	107.3	25,984	26,370	101.5
Jersey City...	Hudson........	93,692	94,252	100.6	26,531	27,349	103.1
Washington...	Dis. of Columbia	83,578	94,046	112.5	23,044	24,683	107.1
Detroit.......	Wayne.........	82,917	83,527	100.7	23,725	24,010	101.2
Louisville....	Jefferson.....	70,685	75,325	106.6	20,868	21,632	103.7
Milwaukee....	Milwaukee.....	69,606	68,931	99.0	19,725	20,244	102.6

By this table of children of the school age it is seen that even in the western cities, where the immigration of adult males brings this sex largely in excess in the general population, the tendency to female births prevails there in common with all Eastern cities. For instance, Cook County, Illinois, (Chicago) has of all ages 98.4 females to the 100 males, while of the school age there are 102.6. St. Louis, Mo., has of all ages but 95.3, and of the school age 104.4 females. Some cities further west numbering less than 100,000 in 1880, where the

immigration was large, show even greater differ-
ences. The county in which Minneapolis lies had
in 1880, of all ages, but 83.6 females to the 100
males, while of the school age, it had 103.9. The
county in which St. Paul lies had then of all ages
but 84.5 females, while of the school age, the enor-
mous difference of 107.4 is presented.

NOTE C.   (*See page* 66.)

Since the publication of the first edition of this
book I have received a number of proofs of this
Physical Law from the observations of others.
They are of the same general character as those of
my own given in the book, which might have been
largely multiplied then, and I do not know that it
would add greatly to the confirmation of this law,
to add others now, as all such observation are
open to the charge, by one who does not accept
the law, that they are specially selected to support
a previously adopted theory, and that others of
opposite character might possibly be found if one
took the trouble to seek.

Such objections can not, however, be so fairly
advanced when confirmation is drawn from one
who deduced his conclusions before the publication of my book, and seemingly without any theory
of sex control in his mind.

Thus Dr. John Stockton-Hough, a long-time investigator of the influences of sex as affecting the
physical conditions of the human race, and the
author of a number of magazine articles of **high**

character on various phases of this subject, in an article in the Philadelphia *Medical Times*, Dec. 27, 1873, reviewing at length some statistics of the State of Michigan, then recently published, bearing on the influence of the foreign born fathers and mothers on the number and sex of children born in the State, ends his article with the following conclusions :

" 1. That foreign born parents are more prolific and have more male children.

" 2. That foreign born mothers with native fathers, are more prolific and have more males.

" 3. From this that the begetting of males requires, in the mother, a more perfect procreative faculty, and the begetting of females requires, in the father, more procreative vigor than the begetting of males."

The tables in my book show that conclusion 1 and 2 above, are true only when the foreign born parents are compared with the weaker native born women in cities. And that it is not the characteristic of foreign birth, but rather the more robust health of these foreign born mothers, a condition found equally if not superior in the native mothers of rural districts, to which a large proportion of male births is due. This indeed seems to be deducible from, if not actually expressed in the 3rd conclusion, above.

For further proofs on this point see Note D.

NOTE D.   (*See page* 70.)

In an interesting series of articles on the " Influence of Sex on the Mother, by John Stockton—Hough M. D., (mentioned in Note C.) in the Obstetrical Journal, N. Y., 1884., he gives a table of 27 observations among families of the British Peerage, and 69 from families in Hyde's Genealogy, made up by himself, showing the intervals of time occurring between the birth of one child and the conception of the next, as this interval was influenced by the sex of the previous and of the later child : Nine months preceding the recorded date of the birth of the later child being taken in all cases as the period of conception.

His record of the first series of 27 from the British Peerage shows the following averages :

When the previous child was a girl	and the next a girl,	the average interval was 5 mos. 23 ds.
	when the next was a boy,	" " 7 mos. 23 ds.
When the previous child was a boy	and the next a girl,	" " 10 mos. 27 ds.
	when the next was a boy.	" " 14 mos. 23 ds.

In the second series of 69 obs.

When the previous child was a girl	and the next was a girl,	the average interval was 14 mos. 19 ds.
	when the next was a boy,	" " 13 mos. 20 ds.
When the previous child was a boy	and the next was a girl,	" " 16 mos. 26 ds.
	when the next was a boy.	" " 19 mos. 26 ds.

Although these observations may need confirmation before they will be accepted as conclusive for all classes and conditions of society, so far as they go they corroborate my conclusion that wives who might otherwise have sons, will, if they have children following each other too frequently, have a succession of girls. In the first series it will be noticed that conceptions that occurred within six months after the birth of a girl resulted in a girl again ; while if the conception was delayed to eight months and over the result was a boy. And when the prior child was a boy additional time was required for recuperating the mother's strength. So that conceptions occurring even within eleven months after, resulted in a girl, and that a recuperation of over fourteen months was then required for the conception of a boy.

The inference here seems a very plain one that the gestation and parturition of a girl is less exhausting to the mother than is that of a boy ; and that it takes her longer to recuperate from the latter than from the former strain on her vitality. Consequently that there is an interval during this recuperation, earlier or later according to the sex of the previous child, when the mother may conceive a daughter while she has yet not strength to conceive a son.

In the second series of observations the long interval elapsing before the next conception shows other influences at work delaying conception. These were probably prolonged lactation, or prudential considerations. Either of these as factors would make the problem more complex, and nat-

urally the influence of the sex of the previous
child in reducing the mother's strength must be
secondary to these more immediate influences.

NOTE E. (*See page* 73.)

The general conclusion that very religious women are often of feeble health is very fully proven
in the work of Mr. Galton on Hereditary Genius,
mentioned in Note A.   He says, chapter XV. ("*On
Divines*") wherein he tabulates statistics from 196
subjects in " *Middleton's Biographies* ": " The fre-
quency with which divines become widowers is a
remarkable fact, especially as they did not usually
marry when young. I account for the early deaths
of their wives on the hypothesis that their constitutions were weak, and my reasons for thinking
so are twofold.   First, a very large proportion of
them died in childbirth, for seven such deaths are
mentioned ; and there is no reason to suppose that
all, or nearly all, that occurred have been recorded
by Middleton. Secondly, it appears that the wives
of the divines were women of great piety. Now it
will be shown a little further on that there is a frequently correlation between an unusually devout
disposition and a weak constitution."

The "further on" contribution to this subject is a
list of 26 of the 196 Divines in Middleton's biographies who were especially feeble in health.   He
gives the particular condition of each of these, re-
marking (page 265) " As regards health the constitutions of most of the Divines was remarkably

bad. It is, I find, very common among scholars to have been infirm in youth, whence partly from inaptitude to join with other boys in their amusements, and partly from unhealthy activity of the brain they take eagerly to bookish pursuits. Speaking broadly there are three eventualities to these young students. They die young; or they strengthen as they grow, retaining their tastes and enabled to indulge them with sustained energy; or they live on in a sickly way. The divines are largely recruited from the sickly portion of these adults. There is an air of invalidism about most religious biographies that also seems, to me, to pervade to some degree the lives in Middleton's collection."

NOTE F.   (*See page* 118.)

N. Y. Tribune December 23, 1884. Extract from an interview with John Glenn, representing Marcus Ward & Co., Irish Linen Paper Manufacturers.

" An experience of twenty-five years in the manufacture of paper has convinced me that climate has much to do with the quality of fine writing papers. * * * Every English workman in an American paper or cotton mill will tell you that *the electricity in the air in this country is much more intense than it is over the water.* So much so indeed that special apparatus is made to correct its influence about a machine." [*Italics mine.*]

NOTE G.   (*See page* 132.)

ADDITIONAL CONCERNING BRUTE ANIMALS.

There are two other theories somewhat current among cattle breeders for controlling the sex of their stock, besides this one of Mr. Fiquet, which may be appropriately considered here.

The most prominent of these is that known as Thury's, in which it is claimed that conception occuring in the early period of the heat, when, as is said, the ovum is yet imperfect, or at least immature, will result in female offspring, while when occurring later along in this period it will result in male offspring. Some breeders tell me they have tried this without any success, while others think they find some proof of its satisfactory working.

According to the physical law I have set forth in this book, it should sometimes be found efficient in producing the desired result, but is not certainly to be relied on. The theory no doubt had its origin in the fact discovered by Huber that if the queen bee has free access to the male at the earliest day she will receive him, five-sixths of the eggs she lays will hatch out females, while if this intercourse with the male be retarded five or six days the eggs she lays all produce males. Assuming that this detention from the male, with her ovaries filled with unimpregnated eggs, tends day by day to increase her sexual desire, such male results are, by the law given in this book, what might be expected.

That this theory of Thury's can not, however, be

of general application is very evident when we consider human beings. Under its action conception occurring almost universally late the children born should be all boys.

Like with the theory of Fiquet, its advocates will have to take the physical character of the parents into consideration, as a weak female in the highest ardor of her heat might be inferior in sexual vigor to the male, and therefore conceive a female, and the strong female at the earlier period be still too ardent for a weak male, and consequently conceive a male.

The other theory, of older date is, that sex alternates with the periods of heat. That if a cow has a bull calf, and conceives again at the first heat after calving her next calf will be a heifer. But if she is passed over to the second period the calf will be a male. And reversely if the prior calf was a heifer it will be followed by a male if she conceives at the first following period, but if passed over to the second it will be a female again. The breeders I have spoken to do not lay much stress on this, yet it has its advocates.

I conclude there may be just this much effect in it. The gestation and birth of a bull calf weakens the cow more than does a heifer calf ; and in this weakened state if she comes in heat promptly the bull is more likely, all others things being equal, to be prepotent, and hence by the physical law I give, she will conceive a heifer. While with the less exhaustion of a heifer calf first she will sooner be strong enough to conceive a male calf. This is in accord with the statistics of the English **Peerage** collected by Mr. Galton (previously mentioned in these notes) which show that conception ordinarily follows sooner after the birth of a girl than after the birth of a boy. And when the next conception proves to be of a boy the interval is still longer than when it is of a girl.

But there is one very strong argument against this alternate theory. If it were true, the puppies in one litter and the pigs of one farrow should be all either male or female. Even if some different rule prevailed with these multiparous females, so that the several ova impregnated at the one time alternated in sex, then the sex should be in equal proportion at each birth. So with twins among cattle. They should either be of one sex uniformly, or always one of each.

It may interest some breeders to know the underlying principle said to be advanced by Mr. Fiquet to account for the results he claims from feeding to control the sex.

It is that low feeding tends to produce a weakened condition in that parent, and that there is some wise law in nature through which efforts are always put forth to sustain in an exceptional way any weakening or debilitating tendency. And with cattle if the bull is artificially weakened this natural law intervenes to increase the bulls in his progeny.

Though this is directly opposed to the evolution theory of "the survival of the fittest" it is certainly **very ingenious, for such a law is apparent in some**

conditions of organic life. The vine throws out
most tendrils where greatest resistance to outside
influence is required. Isolated trees are more
strongly rooted on the side from whence come the
strongest winds. Even with animals, mothers give
more care to their weaker offspring than to the
stronger. Indeed, the whole care that mothers
give to their offspring might come under this law.
With human mothers this attention may be regarded as the promptings of duty, but even with them
it is more an instinct than a prompting of reason.

Shortly after the publication of my book there
appeared in the *Spirit of the Times* [New York,
February 7, 1885,] an interesting communication
from Mr. T. B. Armitage, of N. Y. City, who from
an investigation of horse and farm stock mainly,
had independently arrived at the same conclusion
regarding the influence controlling sex as myself.
The article is too long to copy entire, but is well
worthy of a perusal by those who wish to confirm
my conclusions. His principal points are as observed by himself—that very old stallions or those
which were kept very fat and with such little exercise that they became sluggish got a much larger
proportion of male colts, as large sometimes in
the season as 85 to 95 per cent. with only 10 or 15
per cent. fillies. That on the other hand when the
mares put to a stallion of full average vigor were
old the produce was very largely female.

Another point noticed by him confirms the conclusion I arrived at that the occasional sons of
delicate mothers (those who are having an unduly

large proportion of girls) are generally delicate
in health and often die off before reaching man-
hood. He mentioned "the lady Van Buren of
Orange County, N. Y., a brood mare of very delicate physique who was bred to various horses always producing fillies : and who later being cov-
ered by Rysdyck's Hambletonian in the weakness
of his extreme old age had a colt, Atcheson, who
was so delicate in health that he found an early
grave."

The statistics of Carl Düsing (See Note A.) show
that as the mares increased in number put to the
same stallion in a season there were more male
colts ; this investigation was said to have been
carried to the extent of over a million colts.

It is quite a natural conclusion from this table
that as the number of mares was greater the stallion was more frequently found in a state of sexual
exhaustion from a recent previous service through
which the mares would oftener be pre-potent to
the more frequent determining of male colts.

Per contra, to the conclusion that it requires
the strong mother to produce males while the
weaker one will have females, we have the widely
known experiments of the Frenchman, Professor
Martegoute, some 25 or 30 years ago on a flock of
sheep. The facts he elicited were that the ewes in
the flock giving birth to females were heavier—
fatter—at the time of conception than those that
had males, and at the time of weaning they had
lost a larger proportion of their weight than those
having males. From which the conclusion was

reached that it took more strength to conceive as well as to nourish a female than it did a male. Though the latter part of this conclusion is directly contrary to the experience of all medical men as well as breeders of farm stock, the careful, frequent and regular weighing of each sheep, from first to last, seemed to make such a conclusion the only one that could be reached.

Fortunately, however, the record has it in one little word that throws light on the whole experiment and gives a clue to a wiser conclusion and one in accord with everyday experience. The ewes conceiving the females were heavier—"fatter"—that is the word fortunately added which enlightens us. As is well known strong sexual vigor is inconsistent with fatness. The readiness with which animals after emasculation or spaying take on fat shows this. These fat ewes had less sexual vigor, were less active than the leaner ones, and when coupled with vigorous males, would according to the principle in this book, conceive female offspring. Had they been served by very fat males, or those with a sexual vigor inferior to themselves and conceived male lambs they would no doubt have lost still more weight than they did.

This elaborate experiment is introduced, first, because it is often cited to show it requires strong females to conceive females, and therefore needs to be explained, and secondly as a caution to readers to examine carefully into the data from which conclusions are drawn, even though the compiler has the sounding title of "Professor."

From some actual experiences among my acquaintance owning breeding mares, I am led to the conclusion that in controlling the sex among farm stock, exercise of the muscles carried to even slight fatigue on one side has even a weighter influence in putting that animal, whether the male or the female, for the moment in the inferior sexual condition, than might be inferred from what is said (page 55) in connection with the two sows. The reservoir of nervous or electric energy in the animal from which the sexual energy must be more especially derived at the time of procreation is necessarily worked off for the moment by any active exercise of the muscles. And when a mare or other brute female is driven even a mile or so, though the gait be moderate, some of her muscular energy is worked off, and if she be at once put to the male, without having time to rest, the chances are very great, all other things being equal and he unfatigued, that female offspring will result. If, however, male offspring be desired at the time, giving the male a similar but more vigorous exercise before coupling, during which the female is resting will tend to place him in the inferior sexual condition with better chances for a male. I say this much, though would not have this solely relied on.

I make no apology for presenting the following letters, believing they will be interesting to readers:

" *My Dear Mr. Terry:* A circumstance related to me by the late James B. Richards, A. M., a gentleman well known in Boston and New York, as a teacher of backward and imbecile children for nearly forty years, seems to accord so well with your theory, that I suggest its insertion, as a note in some future edition of your book. It is as follows :

" In the town of Cummington, Mass., two cows belonging to near neighbors, were driven about two miles to a bull. One of the cows was small, plump and handsome; the other was very large, bony and strong. The bull seemed to take a fancy to the small cow and served her three times, and would take no notice of the large cow, though she made every effort to attract his attention. The small cow was then turned into the road and driven home.

" The bull had expended his vigor and satisfied his desire on the small cow, and for an hour paid but little attention to the overtures of the large one ; but he finally served her once only. The result was that the little cow had a heifer calf and the large one twin bull calves. This result was quite unexpected, and always somewhat of a mystery to my informant, as it was to myself, till I read your book, which clearly explains it.

" Even had the smaller cow been ordinarily as strong sexually as the bull her fatigue following the long drive placed her in the inferior condition to the fresh bull; and the heifer calf by your theory would result. The large cow had time to rest and under circumstances conducive to a higher sexual excitement, while the bull's sexual desire as shown

by his delay, was at zero.   Hence in the encounter
she was pre-potent with the result therefrom, in
accordance with your theory that she conceived
the bull calves.   NELSON SIZER."

The past summer a Kentucky gentleman, a
stranger to me, in ordering a copy of "Controlling
Sex in Generation" from the publisher, mentioned
the great interest he felt in the subject, his father
having been largely engaged for years in breeding
Kentucky "short-horn" cattle.   His letter was
brought to my notice, and anxious, as always, to
get practical experiences on this subject I wrote
asking for his.   I append to this note the principal parts of his letter in answer, they must be regarded as very conclusive in confirmation of my
theory.

Mt. Sterling, Ky., 5th July, 1886.
Mr. Samuel H. Terry, New York.

DEAR SIR: I have had but little time to answer
you sooner, or even to read your book carefully
through, but a glance has shown me your theory,
which corresponds entirely with my own, only that
yours is complete and shows careful study and
deep thought which I have never been able to
give to it.

It is not necessary to tell when this theory was
first brought to my notice, or how I gradually reduced and formulated it to this—that the offspring
will be the opposite sex to the parent most passionate at the time of sexual intercourse—which is in
substance the same as your theory.

The part I could not explain satisfactorily was
—why does the most passionate parent impress
the opposite sex on the offspring? If the generative organs are features impressed like other features why not be masculine if the male is the
strongest and most passionate character? I think
you have answered this in a very conclusive and
scientific way. Thanks to you, I now have that
connecting link.

But to the subject. My father was and is a large
stock raiser and breeder of fine cattle, which gave
me a splendid opportunity to observe the condition and study the character of cattle, particularly
with a view to proving my theory about formed
from observation of people. Besides, too, if the
sex of animals could be controlled it would be
valuable knowledge to the farmer, particularly to
a farmer who raised such a class of stock that the
sex would sometimes make thousands of dollars
difference in only one animal.

When my father first went into the cattle business, he bought grade cows simply to breed cattle
for the butchers and naturally wanted all bull
calves, so he could make fat bullocks of them.
And though we thought nothing about controlling the sex then, the fact is the majority were
bulls.

He then had only one bull at a time, to serve a
large number of cows. The bull being a plain
bred one received no attention or extra feed, and
was of course kept very thin by his great use.

Later my father bought thorough-bred short-horn cows, and bred them to bulls of finer pedigree, and strange to say then the majority of the calves were female, just what he wanted. These bulls received extra care and rich feed having separate stalls and lots to exercise in, while the cows were let run out the year round, winter and summer, on grass.

In the last few years as our herd became finer, and somewhat reduced by sales we let the bull that headed the herd stay out in open pasture with a few, perhaps twenty-five, cows. He was, however, kept up at night and fed rich food while the cows were not.

In this plan, somewhat contrary to my expectation, we still had mostly heifer calves, and my observation led me to explain it in this way. I noticed that the bull would court a cow several days before she showed by any outward signs that she was in heat, and would thus take advantage of her before she had much or any passion.[1]

These three ways of handling cattle served the purpose of controlling the sex just as we wanted, but wholly accidental on the part of the breeder.

From previous observations among my friends I had almost arrived at your theory, which was confirmed by the three methods I have mentioned, and fully proven by numerous individual cases I have experimented with. One I will relate.

---

1. A hint for those who would profit by the remarks in Chapter VII. Such conditions work on both sides to the production of female offspring. It is not alone that the female's desire is yet at a low ebb, but through this craving expectation, day by day, the male's sexual desire is wrought up to a higher degree than ordinary.

Several years ago I bought a noted cow, Orphan Nell 5th ; and was told by the herdsman of the last owner that she never gave any outward signs of being in heat, that the only way to tell when she should be served was to refer to the breeding list and see when she was last bred ; which I found to be true.

Though she was a very large cow, weighing 1800 lbs. when I bought her, and looked very vigorous, she never showed the least passion or indicated as other cows do that she was "bulling." So that I could only tell when it was time for her to be in heat by referring to our breeding list or private calf register. If it were time, she would barely submit to the service of the bull when held by a halter. This also was a peculiarity of her produce all females—which I bred for three generations. And not while they were in our herd did they bull as other cows do ; and never did they have a bull calf while I owned that family.

Now you gather what my experience has been, that that animal which is the most passionate produces its opposite sex ; and all other things being equal the animal that is kept in the highest condition by rich food is the most passionate.

In the case of Orphan Nell 5th, it would seem an inherited peculiarity of temperament. But while she was fat when I bought her she never was so again, for she was a great coward and would be imposed on by the other cows, as also would her produce.

One other general case, which makes me believe

in D. D. Fiquet's theory of feeding. Mr. T. C. Anderson of Sideview, Kentucky, a very noted breeder of short-horn cattle has a great reputa- . tion as a heifer breeder. Indeed it is proverbial that Mr. Anderson can get a cow calf almost every time. He is equally noted for keeping his two or three hundred cows in a very poor condition, while he stall-feeds his service bulls, which are always in strong, vigorous condition. Most people attribute Mr. A.'s great run on heifers to his good luck, while I have insisted that it was controlled by the laws of breeding, one of which Mr. A., perhaps unwittingly, complies with. Though what I have hurriedly written is only a general summary of my observations, I trust this is what you want. I believe in this theory, and have for years, and every case I look into and test convinces me of its truth. I think one can breed sex with more certainty with dumb brutes than with people because there are not as many things to take into consideration that serve as exceptions. I believe too in exercising the parent until tired if you want the sex of that parent in the offspring. I think though that if you let the animal rest after the exercise it will serve only to make him or her more vigorous. On the principle that one is very nervous just after a charge from an electric battery, but the nerves become steadier than ever a while after the shock.

Very respectfully,
T. J. BIGSTAFF."

It will be seen that Mr. B.'s experience embraces an example of D. D. Fiquet's theory of low feeding of the male and high feeding of the female when male offspring are wanted, and reversing this when female offspring are desired. And also a phase of the Thury theory, that conception occurring early in the period of heat produces females, and later in that period males. But both these are merged into the more general principle given in my book—that the parent sexually strongest influences the sex to the opposite side.

NOTE H. (*See p.* 147).

CONFIRMATIONS OF THE ELECTRIC THEORY.

The hypothesis that reproduction in all plant and animal life is an electrical process, as set forth in the closing chapter, was purely a deduction from the various conditions which everywhere seemed to surround the reproductive organisms, and accompany the procreative act itself. Not a single pointed fact could be called to mind beyond what is in the book that was of enough weight to raise the hypothesis to the rank of a theory: at least none that would be satisfactory to intelligent readers. It was, of course, hoped that some of these might be led to consider the subject and aid the evolution of the idea by pointing out corroborating circumstances, or that my own further research might elicit some positive proof of its truth. I am pleased to say that I have recently had more light thrown on the subject from the record of

some phenomena observed years ago by scientific
investigators, and which goes very far toward
proving the theory true—in the main, if not in all
its details.

Incidently looking into the subject of Phospho-
resence in Chambers' Encyclopædia I found reference under the heading "Luminosity of Organic
Bodies" to the work of Thomas L. Phipson, Ph.D.,
F.C.S. on "Phosphorescence; or, the Emission of
Light by Minerals, Plants, and Animals." (London, 1870.) In part II., chap. I., p. 69 of this work
accounts are given of observations made by different botanists, from the time of Linnæus down to
more recent years, of flashes of electric light, seen
in the twilight or darkness, playing around the
flowers of several varieties of plants during the
time when the pollen was ripe and falling. These
were seen oftenest in the Nasturtium, Japan Lily,
Marigold, and other flowers of like yellow color.
" In some of the palm variety where the male and
female organisms are enclosed in a spathe, the
rupture of this was at times accompanied by a
loud cracking noise and a spark of light." These
phenomena indicating, as I think, a mutual electric
discharge occurring at the times between the male
and female organs.

In Gray's Botany, Sec. 372, [1] the author dwells at
some length on the evolution of heat in flowers
—a well-known correlation of electricity—saying

1. Structural and Systematic Botany by Asa Gray, M. D., Professor
in Natural History in Harvard University. Edition of 1857.

"it often arises to twenty and even fifty degrees above the surrounding air"—and " *is most striking during the shedding of pollen.*" (Italics mine.) He also remarks on the often peculiar movement of the pollen;—that "It is sometimes projected upon the stigma by transient and other sudden movements, either mechanical as in the Kalmia, or spontaneous and vital as in the Barberry"—where "when excited by a touch on the inner side of the filament it approachês the pistil with a sudden jerk. The object being evidently the dislodgment of the pollen from the anther cells and its projection on the stigma. * * * Anatomical investigation brings to view no peculiarity that might account for these movements." (P. 674)

I think these may be well accounted for, if we accept the increased heat as evidence of an excited electric condition.

Elsewhere he remarks "the pollen in many plants has tubes which while yet in the anther point towards the distant stigma, as in milkweed where the tubes form a tuft or skein." Is there any known influence in nature that will account for this reaching out of the male product of the insensate plant toward the female product other than that of an attraction between the two diverse electric conditions? I can conceive of none.

But returning to Phipson's book ; insect life, as seen in the fire-fly, glow-worm, etc., gives even greater corroboration to this electric theory. This writer has seemingly gathered in his interest-

ing work all the authentic history on the subject.
Before going into the detail of this it is proper
to premise that the lights spoken of by him and
others, as seen in insect life, are uniformly called
phosphorescent, though he mentions that an English chemist, Mr. Thornton Herapath, asserts "the
most delicate analysis did not show the slightest
quantity of phosphorus (as phosphate) in the bodies
of these insects." Later on Mr. Phipson in reviewing the history of the various investigations
on the subject of Phosphorescence in insect life
from 1592 down to the publication of his book,
giving various theories of its origin—none of
which, by the way, refer it to reproduction, says,
p. 178: "Dessaignes concluded that every case of
phosphorescence is closely connected with electricity. Later, M. Becquerel and M. Biot in France
and Professor Henry in America repeating the experiments arrived at the same conclusion."

In passing I may add here that Professor C.
Matteucci in his world-renowned series of lectures in the University of Pisa, 1844, translated by
Dr. J. Pereira, of London, says (Lecture VIII.,
p. 172 of translation): "The phosphorescent matter
of the glow-worm does not present any obvious
trace of phosphorus. Of this fact I have
assured myself, * * * we can no longer regard the
presence of phosphorus as the cause of the light."

These conclusions of scientific investigators
should very definitely settle the matter—that the
so called phosphorescence found in many insects,
though it may continue to be called by that name,

is really electricity, or as least electrical in its luminosity.

Below, I quote some of the more striking observations of the various naturalists mentioned by Phipson, tending to show the connection of this luminosity with the propagation of species.

"Professor Dunneville [Institute of France] writing on this Phosphorescence as seen in the common earth worm (*Lumbricus*) says 'the earliest-known mention of it is by Flaugergues in 1771, '75 and '76, who remarked: The light was emitted principally from that portion of the body in which are situated the organs of reproduction. This peculiarity was also noticed later by the naturalist Bruguiere.' — See *Journal d'Histoire Naturelle*, Vol. II., p. 267."

"Professor Moquin-Tandon member Academy of Sciences [France] says that in connection with M. Sargey he observed this phenomena in the genus *Lumbricus*. 'The light shown was nearly white and resembled a bar of iron heated to a white heat. * * * Each of the Lumbrics was remarked to have a well-developed *Clitellium* which proves the worms were adults and it was their period of coupling. He preserved them several days and observed that the luminous property resided in the sexual swelling or clitellium, and that their phosphorescence ceased immediately after copulation.'"

"Dr. Lallemand being present with a number of other savants at the house of M. Berard, Montpelier, showed them a very curious experiment,

He placed on his hand a female glow-worm and put it out of the window. Very few instants elapsed before a male *Lampyris* flew into the Doctor's hand and coupled with the female. But as soon as the act was completed the light of the female was extinguished completely. This was witnessed by M. M. Berard, Doyes, Dubreil, Balard and Moquin-Tandon." [Page 142.]

The series of lectures by Professor C. Matteucci just referred to, give many interesting facts bearing on this subject, as also generally on electricity as found in animals. And though none of them so specially connect this luminosity with the procreative organism they are all in harmony with my hypothesis. It is impossible to do justice to this special character of these lectures by any short abstract. The reader is referred to the work itself ; the English translation was republished in Philadelphia in 1848.

The investigators of this subject of animal electricity were quite numerous a generation back, and much that they then brought to light seems almost buried from the sight of the present generation. This is not to be wondered at. Their investigations seem so far to have resulted in no practical benefit to humanity, while those made in the line of the arts by which electricity has been brought into a practical use for man's comfort and happiness—in the telegraph, telephone, electric light, etc., have drawn investigators away from the study of the, as yet, more speculative animal electricity.

No excuse is therefore needed for my not having brought these proofs forward at the publication of the first edition. They had almost to be resurrected from the dead past.

A prominent scientific association in this country who have a standing committee or bureau for the especial purpose of giving information to investigators seem not to have been acquainted with them. As when the hint of possible visible electric conditions in plants in connection with reproduction first came to my notice I wrote to the secretary of the bureau of this association, asking if there were any similar phenomena known to the scientific world in animal life ; mentioning also the object for which I sought the information. A few days later I was courteously informed that "Scientists are in possession of no facts bearing on the theory in question."

Doubtless the misleading title "Phosphorescence," under which the discoveries in this electric luminosity has been concealed, had much to do with their ignorance of the details of the books from which later I have derived the information.

The most voluminous writer on this subject is Professor Emil du Bois Reymond, of Berlin, who in 1848—1849 and in 1860 published a series of three large volumes entitled, *Untersuchungen über Thierische Elektricität,* in which the electric lights seen around the flowers of plants at the time of fertilization, and electric phenomena connected with the vital processes of animal life is discussed at great length.

Unfortunately for English-speaking people, this valuable work is yet sealed to them in its original German,—no translation, to my knowledge, ever having been made. Though in the years immediately following the publication of the first two volumes several notices of them appeared in English and American scientific journals from which the general character of the investigations became known to the men living thirty years ago.

Possessing but a very limited knowledge of German I can not attempt giving a resume of Professor du Bois Reymond's investigations. They seem to be in the main similiar to those of Matteucci; and I will content myself with one short passage showing that man is more frequently in the positive electric condition and woman more frequently in the negative electric, as I suggested the probability of, at times of procreation.

After relating the method of some experiments made by Ahrens in 1817 he quotes him as saying: "Die wichtigsten Resultate dieser mehrere Monate hindurch vorgesetzten versuche waren folgende. In der Regel ist die eigenthümliche Elektricität des Menslichen Körpers im gesundheit Zustand positiv. * * * Reizbare menchen von sogenannten sanguinischen Temperamente haben mehr freie Elektricität als trage von sogenannten phlegmatischen Temperamente. Des Abends ist die Menge der Elektricität grösser als zu den anderen Tageszeiten. Geistige Getränke und der dadurch mermehrten kreislauf vermehren die menge der freien Elektricität.

Die Weiber sind öfter als die Männer negativ Elektrisch, doch sind weder die Versuche von Herrn Ahrens—noch von mir, sagt Pfaff,[1] bisher genug vervielfältigt worden, im den Gegensatz der Elektricität des Weiblichen Geschlechts gegen die des Männlichen als Regel aufstellen zu können.

Im Winter sehr durchgekältete korper zeigten erst keine Elektricität, die aber allmählig zum Vorchen kam, so wie die Haut wieder warm wurde.   *   *   *   *   *   *   *

Wahrend der Dauer rheumathischer krankheiten scheint die eigenthümliche Elektricität des Korpers auf Null herabzusinken, und sowie die krankheit weicht allmählig wieder zum Vorchen zu kommen."

The subject is by no means concluded. Indeed our feet are but splashing in the surf of the great sea of vegetable and animal electricity, which, spread broadly and openly before us, awaits and welcomes investigators. As one of these, in a a small way, on an important subject, I will say that I doubt not but the phosphorescence of the ocean so often seen and described, will eventually be shown to proceed from the electric conditions accompanying and bringing together in procreation the myriad male and female inhabitants therein ; and their generative products, as these are ejected and mingle in the water at the seasons of reproduction.

---

[1] These investigations seem mostly to have been by Ahrens, under the direction of Pfaff, who wrote them out.

SUPPLEMENTARY CHAPTER.

It will not be out of place here, at the close of this series of notes, to review some of the more prominent scientific hypotheses of men eminent as writers on this subject of reproduction, especially as some of their conclusions have been cited in print, or in private correspondence as arguments against the physical law of sex set forth in this book. And this present will therefore be an effort to show that this law is really in harmony with the facts on which their hypotheses are based, if not in actual harmony with the hypotheses themselves.

It is claimed to be already proven that in many plants and animals actual fecundation does not take place till two or more days after insemination. That during these days the male germs are slowly finding their way up the fallopian tubes toward the ovum, which may be descending; or with vegetable plants like maize the pollen is travelling along the connecting "silk" at the end of the incipient ear of corn, or through the pollen tube to find the ovule.

This conclusion of a delay was arrived at because if the fallopian tubes be cut or strictured within a day after insemination so the spermatozoon cannot pass up nor the ovum down, or with maize the "silk" cut off close to the ear some time after the pollen is permitted to fall on the exposed ends of it in the ordinary way, neither the ovum in the one case nor the ovule in the other is vitalized. And hence it is charged against my theory that the relative sexual energy of the two parents at the time of insemination can have but little influence on the embryo.

In answer to this I say:

1. That admitting this delay, whatever vigor was imparted to the two, male and female, products at insemination would proportionally inhere at their junction with the same result as if their junction had been immediate, and

2. I will go further than this, and claim that there must at once on insemination be a connection established between the male and female products. Just what this connection is, with our present knowledge, it is not possible to say. But it may reasonably be inferred that with the maize the "silk" forms some electric connection at once between the pollen and the ovule; and in a similar manner that there is some yet invisible chain of communication set up between the spermatozoon in the womb and the ovum in the ovary so that they are at once *en rapport* with each other, to the subsequent exclusion of any other spermatozoon ; even though they may not meet to form a living embryo for a time, during which this connecting line may be cut and their later meeting be prevented.

I come to this conclusion from such facts as the following: It is not an infrequent occurrence in the country, where dogs roam around at liberty, for a bitch to be coupled to two different dogs during the same period of heat, with the result that puppies are found in her subsequent litter so much resembling each dog as to make it quite certain that superfœtation had been accomplished and that each dog is the progenitor of the puppies resembling him.

Unless there is some such immediate connection, as I suggest, between the male germ and the ovum at the moment of the sexual orgasm, which if undisturbed would bring them together, I do not see how the spermatozoon of the second dog could get precedence of those of the first, which from many observations in such cases are shown to be in abundance in the fallopian tubes and cavities of the uterus, living and vigorous, for days after insemination. These certainly have precedence and at the presentation of any fresh ovum would seize on it.

For the foregoing reasons this delay after insemination should not be seriously advanced against my theory.

The next hypothesis to be considered is that known as " Parthenogenesis," which orginated in investigations of the alternate oviparous and viviparous generation of the aphis and some other of the lower orders of animal life. This has been also advanced as an argument against the probability that the male influence on the embryo determines its female condition. As those holding to this hypothesis claim, that, as its name (virgin-generation) would indicate, reproduction can take place without any of the seminal or male germs being supplied to the ovum, and that both male and female offspring are reproduced, capable them · selves subsequently of performing the complete inter-relation of male and female parents to other offspring. From which, if so, it would seem that the male parent must have less influence on the embryo than my theory requires.

As some readers may not be familiar with this peculiar reproduction of the common plant lice (aphides) it is here restated.

In the fall of the year, as winter approaches, males and females are found among these, as among most insects; the former winged and the latter wingless. These unite sexually in the ordinary manner, and the female then lays her eggs on the twigs of the tree or bush which has been her home during the summer. In the spring, with the oncoming of warm weather the eggs hatch out with a few winged males among them but these soon disappear, and all are apparently females. And now comes the strange change in their method of reproduction.

Instead of again having sex intercourse and laying eggs to hatch out as before, each female starts out on her independent account and gives birth to a few living aphides, all females like herself. Without the presence of a winged male she keeps this up all summer, and at short intervals produces other broods, perhaps six to ten times, and all females like herself. Not only this, but her earlier broods soon commence a similar work, and without having ever looked on a male much less felt the embrace of one, they in turn bring forth their broods, all consisting of females. As the summer wanes, however, and winter approaches some of these virgin-born aphides are found to be males. These unite with the females sexually, and the latter lay their eggs as first above-mentioned to carry the race through the icy winter to the next year.

Naturally one who does not go to the very foundational cause of this process will say: " If the female aphis can without intercourse with a male gestate other females, where stands the theory that the male is the prime factor in developing the embryo into a female? If the virgin aphides gave birth to males only it might somewhat accord with your theory, but not so when they produce only females."

Let me, therefore, go into a closer inquiry into this peculiar method of generation, not only for the credit of my book but for the reputation of my sex in this great work of reproduction. For this hypothesis of Parthenogenesis has given rise to some very fanciful and peculiar ideas. The trend of scientific thought, on account of this peculiar aphis generation, has been for some years toward the belief that the male germ is not so important an element in reproduction as it is popularly thought to be. That the poet instead of rhapsodizing on "the embryo slumbering in his sire," should represent it as slumbering in its mother. It is being suggested, if not absolutely held, that the male product acts but a secondary part in reproduction, and instead of being the vitalizing agent of the ovum it is simply nutricial in its character, a something which develops the ovum into a living embryo somewhat like the "royal jelly,"—which fed to the larvæ of the female honey bees develops them into queens, and without which they would be infertile workers. Scientists of highest repute may not use terms quite so pronounced as these, but this is the direction to which their suggestions tend.

There are a large number of scientific and professional writers on this subject, Steenstrup [1], Owen [2], Lubbock [3], Huxley [4], Spencer [5], and many others.

In 1859, when the article "On the Ova and Pseudova of Insects" by Sir John Lubbock appeared, the scientific mind, which for the previous ten or more years had been hesitating at the suggestive hypothesis of a possible reproduction of species without the intervention of the male element, seemed to be reluctantly crystalizing into a firm belief in it,—not so much because it was proven as be-

1. "On the Alternations of Generation in the Aphis," by Professor Joseph J. Sm. Steenstrup, (translation, London, 1845.)
2. "On Parthenogenesis, or the successive production of Procreating Individuals from a single Ovum," by Professor Richard Owen, London, 1849.
3. "A Record of Observation on the Habits of Ants, Bees and Wasps," by Sir John Lubbock. Also various articles by him in the Royal Society's Philosophical Transactions between 1850 and 1860, notably one in 1859, entitled "On the Ova and Pseudova of Insects."
4. "On the Agamic Reproduction, etc., of the Aphis," Thomas H. Huxley.
5. "Principles of Biology," Herbert Spencer.

cause there seemed no other sufficient hypothesis
which could account for the observed alternate
generation. And Sir John in the opening of that
article fairly states the then condition of the subject in the scientific world,—as follows : "In his
celebrated work on 'Alternations of Generations' if
Professor Steenstrup has not succeeded in explaining the phenomena of asexual reproduction
he has at least the great merit of having brought
together many interesting observations, the relations of which had remained unrecognized up to
his time. The value of his suggestions is well-
shown by the number of memoirs which in the
last few years have appeared on this subject, and
by their having produced a discussion in which
almost every naturalist has taken a part. It is,
however, perhaps not going too far to say as yet no
satisfactory explanation of the phenomena has been
suggested, and that we are now just as far from
knowing as we were twenty years ago, what are
the different conditions under which some eggs
remain undeveloped unless they are brought under
the influence of the spermatozoa, while others contain within themselves the power of producing
young without the necessity of any external stim-
ulus."

Perhaps Prof. Owen stands out most prominently
in this investigation, at least he seems to have
given more than others the particular direction
before mentioned in his explanations of these
strange phenomena, and is oftenest quoted, or was
in the first few years, as an authority to sustain

them. The criticisms herein made on this hypothesis of Partheno or Agamic generation will therefore be more especially confined to the views he presents.

He says : " This peculiar alternate generation of the Aphis was first noted by Bonnet, in his *Traite d' Insectology*, 1745, and was then received with great incredulity. Subsequent investigators, however, found it to be true. Reaumur suggested that these so-called females are androgynous." Owen says : " This does not explain it, as there is no trace of the dual sex appendages to be found in these female Aphides ; the larval germs are perceptible even in the embryo aphis before the formation of any receptacles for them."

Further along Professor Owen concludes that : " The original seminal product is in some manner retained in sufficient quantity through these successive aphis generations to fertilize the later ova, or rather embryotic masses which if unfertilized would develop into ova, and from this fertilization viviparous births successively proceed till the germ mass is exhausted, when ova are developed in some of the animals, and seminal organs and germ masses in others to meet together and fertilize the eggs to be laid."

This is a very ingenious speculation if it is not a conclusion acceptable to others. It fails in not giving a general law by which the change from one method of generation to the other is secured· Indeed, this which he calls his conclusion is not in accord with ideas he presents on a collateral subject, as follows: "The first step in the formation of the bud in the Hydra is the multiplication of cells which also are aggregated together as in the ovum. * * The mutual contact of these delicate cells suffices for mutual action and re-action on one another ; excess of nutriment or other quality in one is balanced by the endosmotic action of neighboring cells. * * The young Hydra from the bud is identical in organic structure and character with that which comes from the ovum, and *when the effects of organic development are the same their efficient causes can not be altogether distinct; only the non-essential process may be subject to variation.*" [Italics mine.]

The difficulty has been that these first conclusions, or as they might better be called "speculations" have been taken by other writers as too positive, and so quoted. Even so great a light in science as Huxley follows this lead, and while disputing Owen's explanation of a retained germ mass continued from the original fecundations, through the later virgin generations (which he finds may be extended indefinitely), as so near an impossibility as to be altogether improbable, he goes a far way on towards the idea that the male product in generation is not of high importance ; in some cases at least. Closing his first paper as follows : "Time was when the difficulty of the physiologist lay in understanding reproduction without sexual process. At the present day it seems to me the problem is reversed, and that the question before us is : Why is sexual union necessary? Far from seeking for an explanation of the phenomena of gemmation in the transmitted influence of the spermatozoon [Owen's suggestion], the philosopher acquainted with the existing state of science will seek, in the laws which govern gemmation, for an explanation of the spermatic influence."

So, too, Herbert Spencer, though he does not expressly say so, his language leads to the conclusion that he regards the male element in reproduction of minor importance, and that in some of the lower orders of animal generation it may be entirely dispensed with.

See section 79 of "Principles of Biology," where, after concluding that agamogenesis [1] is the normal method of reproduction among aphides and some others of the low orders of animal life, he considers the question, "Why does gamogenesis [2] recur?" After a considerable discussion of this, he ends by saying : "The above induction is an approximate answer to the question, *When* does gamogenesis recur? but not to the question which was propounded— *Why* does gamogenesis recur? *Why* can not multiplication be carried on in all cases as it is in many cases by agamogenesis? As already said, biological science is not yet advanced enough to reply."

He gives, however, what he calls a "hypothetical answer" at some length ;—too long to be repeated here, but suggests no governing law of nature to

1. Generation without the male.
2. Generation by union with the male.

satisfy the "Why?" his closing remarks on this
being, "Similarly with the aphis. Living on sap
sucked through its proboscis from tender shoots
of leaves, and able to take in but a very small
quantity in a given time, this creature's race is
more likely to be preserved by a rapid asexual propagation of small individuals, which disperse
themselves over a wide area of nutrition than it
would be did the individual growth continue so
as to produce large individuals multiplying sexually.   While at the same time we see that when
autumnal cold and diminishing supply of sap put
a check to growth, the recurrence of gamogenesis
and production of fertilized ova that remain dormant through the winter is more favorable to the
preservation of the race than would be a further
continuance of agamogenesis." Vol. 1, page 236.

It seems to me that this "hypothetical answer"
is defective as an explanation. It is not apparent
from it how the effects can result from the causes
named. The law of nature is that a favorable
environment, such as the aphis has in summer, increases reproduction of the largest and best types
and not of the inferior. The largest number and
best-developed grains of corn are found on stalks
where the conditions of plant life are most favorable.

But even accepting this hypothesis that in the
case of the aphis an abundance of plant food
in midsummer induces an inferior type, and the
destitution in late autumn a superior or higher
type, I fail to see how these effects can react back

—or even collaterally, to induce in the one case agamogenesis, and in the other gamogenesis. With all due respect for the opinions of one ranking so high in the scientific world as does Mr. Spencer, I believe that the explanation of this peculiar alternate generation I shall give a few pages forward, is a more rational one.

But returning to Owen. To show how inconclusive his impressions were I quote some remarks found in a footnote in his book where he reviews the conclusions of Ehrenberg on this subject ; and these contain I believe what will eventually prove to be the true solution of this mystery of parthenogenetic reproduction ; though those who have accepted him as authority on this subject, seem generally to have preferably adopted his novel and less probable suggestion of a virgin generation. His words in that note are: "There is a development or secretion of the germ mass in the body of these female aphides as well as of the ovarian mass, so that in the earliest or incipient condition self-fertilization is made and the embryo larva instead of developing through an egg is nourished in a uterus and born alive."

That this is the correct view of this peculiar aphis generation and as such that it is in harmony with my theory I shall now endeavor to establish.

In seeking for an explanation of any mysterious operation in nature it is wise to search in realms we know something of already. There may be a fascination in explorations into unknown fields through which none have been, and no landmarks

are found, but before we go there we should certainly fully investigate the familiar places. We
have learned something of the method of reproduction in animal and vegetable life as it is accomplished through the united agency of the male and
female sexual productions, whether these be found
in separate individuals of the species as in most
animals, or united in one as in most plants. And
it is in this line we should look for a solution of
the mystery of so-called parthenogenesis.

It is not so many years ago when very large
numbers of the lower orders of plant and animal
life were classed as cryptogamous. Year by year,
however, as investigators were more observant the
circle of these has narrowed. Proper male and
female organisms were found to exist in many, indeed in most of them, either in separate individuals
of a species, or united in one. To such an extent
has this research been carried that physiologists
confidently look forward to the time when all these
secret conditions of reproduction will stand revealed and cryptogamy be a classification of the past.

This being so, when in the further investigation
of these orders we find in some the proper female
organism performing its special work, while the
male organism and its direct influence in fecundation yet remain hidden from us, is it not unphilo-
sophic to hastily conclude it does not exist here?
That in these few cases the order of nature is chang-
ed and procreation can take place without the
male element?

Such has not been the course in other fields.

When some hitherto unseen perturbation is observed in a planet's motion, or an unconformable rock strata is found; the astronomer and the geologist strive to find solutions of the mystery in harmony with what they have learned of nature's orderly laws, and not in erratic deviations therefrom.

Between the lower orders where both male and female products are secreted by one individual of the species and self-fertilization is the rule, and the higher orders where the ovary with its ovum develops in one individual and the testes with their seminal products are found in another, and the two have to join in the reproduction of their species, there must necessarily be some dividing line where both sides meet, and where the difference between them is small. This is not merely probable but may be regarded as absolutely certain. As is often said, "Nature does not advance by leaps," the evolution from the lower to the higher is by almost imperceptible degrees. Or as more scientifically stated by Dr. T. Williams, an eminent British naturalist when, commenting on some peculiarly different conditions of the reproductive organs found in some species of annelids which seemed then to him unexplainable, he says : "They must in the very nature of things be subject to the governance of some Morphological law of more or less extended application. Abrupt transitions and exceptions are impossible." Indeed, in

the article this extract is quoted from ; there is
much to sustain this view of an orderly, if sometimes irregular progression, from the lowest cryp-
togamic reproduction by one individual which is
both male and female, up through various intermediate and irregular conditions not unlike the
aphis generations, till the highest bisexual reproduction is reached, all within the family of the
Annelids alone.

It is a reasonable conclusion, and one warranted
by the circumstances, that the aphides with their
peculiar alternate generation are vibrating on this
dividing line. That in the heat of summer with
abundance of food each individual is crypto-
hermaphrodite, secreting not only the ovarian
mass but also the germ mass, which, at some early
period in its existence before cognizable by the
microscope has its action on the ovarian mass
according to one of Professor Owen's suggestions,
and that this method of propagation continues
while abundance of warmth and food is at hand.
Though the individual thus producing the living
aphides is apparently by this parturition a female,
it is really both male and female. Professor Owen
says : " The number of virgin broods is limited by
the length of the season ":—it is therefore probable
that later, as colder weather approaches and the
supply of food is less, or less nourishing, the
individual aphis has not sufficient vitality to

1. Researches on the Structure and Morphology of the Reproductive
Organs of the Annelids. By Dr. T. Williams. Royal Society's Philosophical Transaction. London, 1858.

secrete both sex products. Some yet unknown influences disabling the organism which secretes the male germs in some individuals, and that secreting the ovarian mass in others ; so that some become more especially male, others more especially female, through the union of which oviparous generation ensues ; and eggs are laid to preserve the species over the cold winter.

That this, as a cause, answers Spencer's question, *Why* does gamogenesis recur? is, I think, fairly proven by the fact, instanced by Huxley, that as many as thirty continuous generations of these parthenogenetic females have been obtained by supplying favorable conditions of warmth and nutriment, and from which Huxley concludes their existence might in this way be extended indefinitely.

I will even go so far in this direction as to suggest to those who have the time and the love for the microscopic investigation, the probability that this mysterious disappearance of the winged, or supposed exclusively male, aphides that are hatched out in the spring from the eggs laid the previous autumn is that they, too, soon develop the dormant feminine organs, cast their wings and are transformed into self-fertilizing viviparous aphides, in which condition they lose their identity as males, and are undistinguishable from those at first wingless. And though both may then be called females because parturient they are really androgynous.

The foregoing explanation of the alternate aphis generation will, I conceive, be found in keeping with known facts in the generation of others of these lower orders. The common angle worm(Lumbricus) is represented as hermaphrodite each individual being capable of self-fertilization, and yet for reproduction ordinarily two unite. If so, it is not improbable that in an unfavorable environment the one filling one part in the reproductive act might be incapacitated to fill contemporaneously the other part. So that if the existence of the individual was of some years duration it would become exclusively a male or a female as in the higher order of animals. And yet that under very favorable conditions an occasional individual might still be found able to act in both capacities, or perhaps (as I suggest occurs with the aphides,) continue to be able to produce self-fertilization.

That such must have been the vibrating course of evolution from the bisexual type to the unisexual will, I think, be conceded by every intelligent man. And in saying this, it is immaterial whether the intelligent man regards the vivifying influence as derived from a quickening power inherent in inorganic matter, or from the fiat of an Omniscient being. For Omniscience must bring the successive creations up through the line of least resistance. Any other more difficult line would show a want of wisdom incompatible with omniscience.

Accepting the foregoing reasonable explanation

of the peculiar aphis generation there is nothing in my book inconsistent with so-called parthenogenesis. Indeed the physical law there announced is capable of explaining this continuous birth of parturient aphides during the summer. In the heat of summer with abundance of food the individual aphis before oviparous has enough vitality to secrete germ masses also, and secretes these in such abundance or with such energy that they, according to the law in my book control in some measure the sex of the embryo aphides, individuals capable of parturition only being brought forth. As the environment becomes less favorable on the approach of winter this added power of secreting the germ mass declines, less of it would be secreted or it would have less energy, so that the female side would sometimes preponderate with the result that according to my theory some with a preponderance of the male element would be born.

Parthenogenesis as interpreted by some scientists, and as the name would really indicate, is certainly inconsistent with all we know of nature's laws in reproduction.

A hypothesis which relegates the male not only to a subordinate condition in the generative process, but actually makes him a nonentity is at variance with all our experience. His marked and super-eminent influence on the size, vigor, action, appearance, temper, and intelligence of the offspring, especially as seen in hybrids, is too pronounced to permit such a hypothesis where any other is tenable.

Darwin in his " Variation of Animals and Plants " cites several instances showing the male parent's influence on the offspring; as that, out of twenty-three kittens, the progeny of a tailless male Manx cat and ordinary females, seventeen were destitute of tails. That all the get of a hairy, goat-like ram from the Cape of Good Hope crossed with ewes of twelve other breeds closely resembled the ram; while the ewes of this begetting put to a merino ram all produced lambs closely resembling the merino.

Even more impossible does this supposed subordinate condition of the male seem to be when we regard the great influence he can exert on the mother herself through the gestation in her womb of the foetus of his begetting. It being well-known that the peculiarities of the male parent of the first issue of a female are often impressed on the later issue by a different male. Mares once bred to a jack, when afterwards bred to stallions almost invariably show indications of the jack in their later foals. Pure bred bitches if they once bring forth a litter of mongrel whelps are esteemed useless for producing a pure breed of pups afterward. They may do so but the chances are that some will have the marks of the prior mongrels.

In vegetable life, where the process of fertilization of the ovule can be more closely observed than in animals, botanists say the pollen germ when it impregnates the ovule apparently sinks into it, becoming as it were rooted therein and grows visibly in that condition, gradually absorbing to itself the ovule and virtually taking its place.[1]

A gentleman of my acquaintance mentions a circumstance occurring under his own observation showing this influence of the male pollen in vegetable life on the plant producing the ovule.

In his father's grounds side by side were two rose-bushes one bearing pure white, the other dark red, almost black roses. The pollen from the dark roses wafted over the white, soon so impregnated the bush that it bore mottled roses, and though subsequently the white rose bush was transplanted to a distant part of the grounds this influence was continued. Here the influence on the seeds themselves of the white rose was unknown as none were sown, but it must have been through the seed that the bush itself became so impregnated with the characteristics of the dark rose that it influenced later blossomings fertilized by its own pollen.

I have now in my possession a fair-looking tomato, or to all outward appearance one; red, regularly lobed, and conformable in shape to an ordinary tomato, which grew the past summer on the bush of an egg-plant. It seems to have more

---

1. This was the teaching thirty years ago, but in more recent years seemingly, without any new discoveries, the conditions are in some botanical works interpreted differently, and in accordance with the supposed new light from parthenogenesis.

of the solidity, though, of an egg-plant than of a tomato. It is doubtless a hybrid from the pollen of the tomato, taking exactly the form of the male parent. Well-formed grains of maize have been found growing on the exposed leaves of a certain grass growing beside a field of maize ; doubtless derived from grains of the maize pollen falling on some favoring matrix on the leaf where it was able to attach itself and grow to be an independent grain. Well-formed acorns have been found growing on a pine tree, no doubt a similar result of the lodgment there of the pollen from an oak in such favoring circumstances that it could grow.

The influence of the pollen, the male element of plants, on hybrids is also a remarkable indication of its preponderating importance in reproduction. Gray's Botany, referring to hybrids, says, " The sterility of hybrids is owing to the impotence of the stamens which perfect no pollen, and most such hybrids may be fertilized by the pollen of one or the other parent. *Then the offspring, either in the first or second generation, reverts to the fertilizing species.*" [Italics mine.] Such facts as I have here presented are too plentiful, and show too forcibly the independence and over-whelming influence of the male germ to allow it to be dispensed with, or even relegated to a secondary position in reproduction.

One of the reasons advanced for regarding the female product as the essential one, and the male as merely nutricial, is that the ovule has been sometimes noticed as going through only a part of its incipient change when one male germ is supplied and then stopping there as if it had exhausted all the nutriment from that germ; going on in its change when a fresh germ was supplied, and thus till three or perhaps more were supplied, when the process went on to completion. It seemed, therefore, as if the ovule halted for lack of proper nutriment which the male germs one after another supplied. But this is just as reasonably explained on the hypothesis that the male germs were weak; the first, second and third unable to secure a permanent hold died off, yet each in its turn better fitted the ovule for its successor. In a somewhat similar way we may explain the oft-mentioned fact that when a male germ of a different incompatible species finds a lodgment on the ovule it is not so thoroughly adherent, but that it may be forced aside by a germ of the same species as the ovule applied even a day later. This excluding process may be likened to a brood of very young animals nursing a foster mother of a different species, as young puppies nursed by a female cat. They might get along tolerably well when not interfered with, but if a sufficient number of young kittens were supplied to require all the teats, the readiness with which these would find the teats as compared with the difficulties the puppies would have, would soon leave the puppies to starve.

I come to the conclusion that there is really no difficulty in giving an intelligent explanation, in accordance with our old-time observations of the superior influence of the male in the office of re-production, or at any rate his equal office, which is all I ask for, to any of these peculiar types of gen-eration, so as to bring them into conformity with nature's well-known laws as well as with the physical law presented in my book. It requires, however, that "suggestive hypotheses" shall not be taken as proven when others quite as reasonable, and in accordance with prior known facts can be ad-vanced.

For example a great mystery is made of the fact that the queen bee produces only drone eggs if fertilization by the drone is delayed till after the twentieth day from her transformation into the insect form, while if her union with the drone takes place at the earliest day she will receive him, about the fifteenth after, five-sixth of the eggs hatch females.

Huber "thinks it is because only an imperfect impregnation is operated at the late date." Jardine[1] repeating these observations says, "It is a great mystery which we are unable to solve." Von Siebold[2] says "Only the queens' and workers' eggs are fertilized by sperm cells stored in the *receptacu-lum seminis*, and these she can fertilize at will by muscles guarding the duct. Drone eggs are laid by unfertilized queens and in some cases by work-

---

1. Naturalist Library Vol. XXXIV, H. G. Bohn, London. Edited by Sir Wm. Jardine, F. R. S.
2. Quoted in Packard's Guide to Insects p. 118, H. Holt, N. Y., 1880.

ers." Other and different views might be added but these are the principle ones held, and are certainly contradictory enough to puzzle the student who seeks in these writers for the truth.

On this subject the article " Bees " in the " Encyclopedia Brittanica " is a mass of contradictions. In one place one opinion, in another a radically different one, and no attempt is made to harmonize them. The final decision there on the disputed question follows Von Siebold's views, and is made without giving any reason for the preference.

But those who will carefully read the literature on this subject, and compare Huber's extensive and careful experiments and researches with the comparatively superficial investigations of the others who adopt different views, will not fail to be convinced that impregnation of the queen bee is always a condition precedent to the laying of fertile eggs.

All the writers unite in saying the cause of her laying at times all drone eggs, originates in the delay in her fertilization. And I suggest that the law governing sex as given by me fully governs in this case, as the following circumstance mentioned by Huber tends to prove. He says " When the queen bee is detained from the drone she becomes agitated, stirring up even the workers." Now this agitation I conceive to be excessive sexual ardor from the retention in her ovaries of the unimpregnated eggs. In this excited condition, when the lazy drone finally unites with her, she is so prepotent that her influence controls the progeny, and they are all males. I do not know that it has ever been tried but according to this "physical law" the proportion of drone eggs should increase day by day after the fifteenth as she is detained from the drone. That is with a union on the 16th day there should be a larger proportion of males, on the 17th still more, 18th yet a greater proportion, and so on till the 21st when all are males.

This "stirring up of the workers" as mentioned by Huber is an odd and curious confirmation of the idea I present that it is the sexual ardor of the queen that agitates her. A similar condition of agitation is found among cows in a herd when one of them is in heat and is not allowed access to the male. Many of them will be so greatly excited, if the matter is prolonged, as to quite reduce their flow of milk for the day. It would be a curious if not instructive subject of study to learn the cause of this and how far it extends among other species of animal life. Might it be that there is a dormant element of the opposite sex still existing in both sexes? A trace of the condition when in the evolution of species all were androgynous as referred to a few pages back.

The fact that females with defective ovaries take on many masculine traits,[1] while emasculated males become effeminate in appearance would indicate

1. See the case of a hen in Darwin's "Variation of Animals and Plants' that acquired many of the outward signs of the cock.

something of this kind.   Such feminine appendages as the paps in man might be accounted for in this way.

Still another modern hypothesis mentioned as in the way more particularly of the acceptance of my electrical hypothesis is "Pangenesis," though I can not see in what particular my ideas are at variance with it.   In the first place it may be said of pangenesis that Mr. Darwin did not put it forward as established or proven.   He offered it, as he said, merely as a "suggestive hypothesis."   Several eminent scientists have vigorously criticised it,[1] and shown its probable defective character as a law governing the transmission of parental characteristics to offspring, and it was not warmly defended from these attacks by its author in his lifetime. So that it should not be considered so well-sustained as to stand especially in the way of any other theory, particularly one relating to another branch of the subject.   The suggestion of the vital gemmules in this hypothesis, as transmitted from parents to offspring, seems not to relate so much to the law of generation itself, as to the method of conveying physical qualities through the generative act.   It may be likened to an inquiry directed to the manner in which electricity is conducted or

1. "In more senses than one Mr. Darwin has drawn heavily upon the scientific tolerance of his age.   He has drawn heavily on *time* in his development of species, and he has drawn adventurously upon *matter* in his theory of Pangenesis.   According to this theory a germ already microscopic is a world of minor germs.   Not only is the organism as a whole wrapped up in the germ but every organ of the organism has there its special seed."   Prof. John Tyndall, L. L. D., F. R. S.   In his lecture "*On the Scientific use of the Imagination.*"

passes from place to place along a wire, rather than into the nature of electricity itself. Whatever pangenesis may eventually become, in its present status my theory that the male influence controls the embryo to the opposite sex, and *vice versa* the female to the male sex, and this probably through some electric action at the time of conception, is not opposed to it. We have but to stretch Mr. Darwin's " suggestive hypothesis " a step further and suppose these multitudinous gemmules gathered from every part of the body of both parents and meeting in the embryo are of diverse electric conditions and that in the conjunction there, the most intensely electric, controls the sex to the opposite side. If the father presents these the offspring will be female, if the mother then male. In reality it would amount to about this requirement from pangenesis. If, in the gemmules proper to form the sex character, segmentation begins at the negative end and growth is inward the embryo would be female, if at the positive end and the growth is outward then the development is into a male. The impulse to grow in one direction or the other being imparted by diversely influencing electric conditions. Such positive and negative conditions are known to exist in the cells of the lowest orders of animal existence during reproduction by gemmation. As previously mentioned in **Note H.** many such electrical conditions are seen to exist in the males and females of the so-called phosphorescent insects at times of reproduction. These conditions are also shown to exist in men and women, and though investigation has not gone far enough with them to clearly show that they are connected with reproduction, as with the phosphorescent insects; the inferential proof from these insects, as well as the electric character of the generative act itself, go far towards making my hypothesis a reasonable one.

www.ingramcontent.com/pod-product-compliance
Lightning Source LLC
Chambersburg PA
CBHW021706210326
41599CB00013B/1537